THE AMBER BOOK

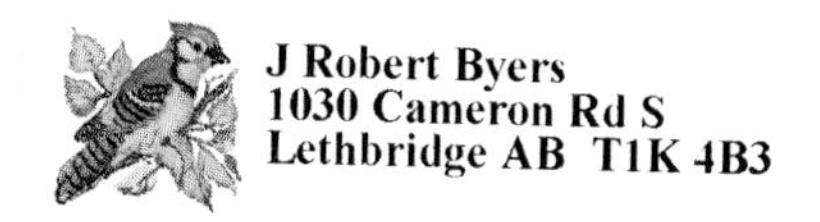

THE AMBER BOOK

by

ÅKE DAHLSTRÖM

and

Leif Brost

Translated by

Jonas Leijonhufvud

Publisher's note: The original Swedish text has been revised slightly in the English-language edition to incorporate minor changes by the authors and to address a wider (international) audience.

English translation published by Geoscience Press, Inc., Tucson, AZ

Library of Congress Catalog Card Number: 96-077632
ISBN 0-945005-23-7

First published in 1995 under the title Stenen som flyter och brinner
by Norsteds Förlag, Stockholm

Printed in Sweden
10 9 8 7 6 5 4 3 2 1

Publisher's Cataloging in Publication: ***(Prepared by Quality Books Inc.)***
Dahlström, Åke, 1922–
The amber book / Åke Dahlström and Leif Brost; translated by Jonas Leijonhufvud.
p. cm.
Revised and expanded translation of: Stenen som flyter och brinner. Stockholm, 1995.
ISBN 0-945005-23-7

1. Amber. 2. Amber art objects. I. Brost, Leif. II. Dahlström, Åke, 1922–. Stenen som flyter och brinner. III. Title.
QE391.A5D34 1996 553.2'9 QBI96-40177

Trade distribution by Mountain Press Publishing Company
P.O. Box 2399, Missoula, MT 59806
1-800-234-5308

The first general work on amber was written by P. J. Hartmann in 1677.

PUBLISHER'S FOREWORD

Amber has been one humankind's favorite decorations and personal adornments throughout the ages. People have always been attracted to and fascinated by these beautiful fossil resins that may contain trapped insects and plants from an earlier geologic time.

This book explains how amber is formed and describes its geologic setting in the Baltic region and elsewhere. The authors explore the role amber plays in the folklore of many cultures, and provide practical advice on how to care for amber jewelry, how to use it for lapidary purposes, and how to detect fake amber. Profusely illustrated, THE AMBER BOOK is a basic source of knowledge for anyone who appreciates amber, including jewelry buyers and art collectors.

Special thanks are due to the translator, Jonas Leijonhufvud, editor Jeffrey Lockridge, and technical consultant Pat Craig.

Amber weighs a bit more than ordinary water. But if salt is mixed into the water, the clear, compact pieces will hover close to the bottom, while the opaque stones, which contain microscopic gas bubbles, will float at the surface. The specific gravity of most amber falls between 1.05 and 1.09.

AMBER FACTS

Data are for succinite or Baltic Sea amber.

Origin: organic (fossilized resin)
Age: 20–50 million years old
Structure: amorphous
Surface fracture: jagged
Streak: whitish yellow
Chemical formula: $C_{12}H_2O$ *or* $C_{19}H_{16}O$
Hardness: 2.0–2.5 on Mohs' scale
Specific gravity: 1.050–1.096
Refraction index: 1.54
Fluorescence: blue-white in longwave UV; green in shortwave UV
Melting point: 480–720°F (250–380°C)
Solubility: up to 25% in alcohol
Content: 70% resin acids, 3–8% succinic acid, and 4–10% water

Even in our time something magical rests within amber, and many carry the stone as an amulet or talisman. The baptismal gift of an amber heart is seen not only as a decorative piece of jewelry, but also as a protector against evil and disease.

Amber has fascinated mankind for thousands of years and has often been attributed divine origins and supernatural powers. It has luster, depth, and substance. It is warm, light, and pleasing to the skin. Energy borrowed from the sun is at work inside it. Amber contains fire, electricity, and perhaps other powers that dead matter does not possess. For amber is born from life; it is a concentrate, the only palpable remnant of the vast forests that once covered northern Europe. Time has stopped inside it.

A sixteenth-century Polish author once wrote: "Amber makes the heart light, keeps bad luck away, and courage strong. It is the remedy of the soul."

Fossilized resin

Just like us, trees bleed when they are injured. Fir and pine trees release a sticky, viscous substance called "resin." In the same way that blood coagulates to protect a wound, the resin solidifies. Hardened drops and streaks can often be seen along the trunks of trees. Soft resin can be used commercially to produce turpentine and, as many know and some have tried, chewed as a gum with disinfecting properties. Soft resin can be used to protect scabbed sores, if applied to the skin, preventing infectious bacteria and fungi from attaching to the wound. The ancient Egyptians did not understand this process, but they knew of its effects when they used resin in their embalming procedures. The ancient Greeks found that dissolving a little resin in the wine made it keep better. This knowledge is still put to use in the making of retsina wines, whose distinctive taste comes from the resin of the Aleppo pine. But resin is also the raw material of amber. During the transformation process, the sap and oil content of the resin is reduced. It is a process that, so far, only nature can manage, and it takes time—millions of years. So don't count on finding amber during a walk in the woods.

The raw material for most of the amber in the Baltic area was produced by conifers around 20–50 million years ago. At least that is what scientists believe. Strictly speaking, it is an informed guess rather than an absolute truth, although the evidence in this case is exceedingly strong.

If we go back these 20–50 million years in time, we come to the epochs of the Tertiary period that the geologists call Oligocene and Eocene. During these epochs both conifers and deciduous trees grew in Scandinavia and in the mountainous mainland that is now beneath the Baltic sea. The trees back then were quite similar to those of today. Of primary significance are a few conifers, most likely ancestors of the pine, commonly classified as *Pinus succinifera*. The species name and *succinite*, the scientific term used to describe the form of amber that is considered authentic (and has always been the primary form involved in trade), are both derived from the Latin, meaning "juice stone."

The Oligocene epoch was a tumultuous time on earth. The Alps and the Himalayas were formed, the climate grew colder, the dinosaurs died out, and the era of the mammals began. Most mammals were still small. Rodents such as rats, rabbits, and squirrels evolved as well as little horses no larger than the German shepherds of today. But large species developed as well. Some rhinoceroses, for instance, grew to hulking proportions, becoming perhaps the largest land mammals of all time.

Because the forests were still pristine in this prehuman era, the resin that was produced when the trees were damaged stayed on the bark as the trees died and decomposed. Over millions of years, this resin solidified and turned into stone. A vast destruction of forests occurred, some scientists theorize, during the Oligocene epoch when the Baltic forestlands subsided and the seas washed in over large areas. It began gradually, as salt water first seeped in to kill the trees, and it ended drastically, when the natural embankments burst and collapsed. Large streams flowed through the areas and carried the light amber with them. Because of the currents' direction, most amber ended up in the Baltic countries and eastern Prussia.

The kauri tree (Araucaria family)

Today most reserves of succinite amber are located within the East Baltic region of Samland between Kurisches and Frisches Haff, close to present-day Kaliningrad (formerly Königsberg). Here large amounts of amber are contained in what is known as the "blue earth." Each cubic yard of this earth, which lies partly below sea level, is thought to contain around 2.5 pounds of amber (1.5 kilograms per cubic meter). That's quite a supply of amber when you consider that the blue earth, which lies beneath 100 feet (30 m) of earth and sand, averages 30 feet (9 m) thick and covers up to 200 square miles (500 km^2) of

Today we aren't sure whether the "amber tree" was a species of pine. Recent research, employing spectral analysis, points to a closer relation with the Araucaria family, which is composed of such trees as the monkey puzzle tree, the Norfolk Island pine, and the kauri pine of New Zealand. Such trees may very well have grown in northern Europe along with a few other exotic species. Pollen analyses indicate this. Continued research may provide more insight.

Thanks to amber, we know what the insects and spiders of millions of years ago looked like. We can conclude that the mosquitoes that filled the forests in prehistoric times didn't look much different from those that torment us today. Insects have been trapped in resin, encased in airtight capsules, and preserved for the afterworld. A better way of embalming is hard to find. Such preservation, enabling a detailed three-dimensional study of the animal's exterior, is an invaluable asset for the researchers. Amber provides a window to life in a world 15–320 million years old.

land area. Blue earth is actually not blue, but black or gray with traces of green. It is composed mainly of clay, not blue clay, but rather glauconite-rich clay. Blue clay amber reserves also exist in some parts of Poland, albeit deep underwater, where they are hard to reach.

Amber has also been discovered on certain islands in the southern part of the Baltic Sea. Usedom Island, which once belonged to Sweden, is said to contain a reserve of substantial value.

Amber is found in areas where various upheavals, such as volcanic eruptions, rift formations, faults, and other vertical and horizontal movements, have affected the earth's crust during the millions of years since amber originated. In many of these cases, the earth's strata have been overturned.

The ice ages have also played a part in the movement of amber, and for this reason amber can be found as far south as Central Europe. Much of the amber that the ice brought with it was transported to the sea through creeks and

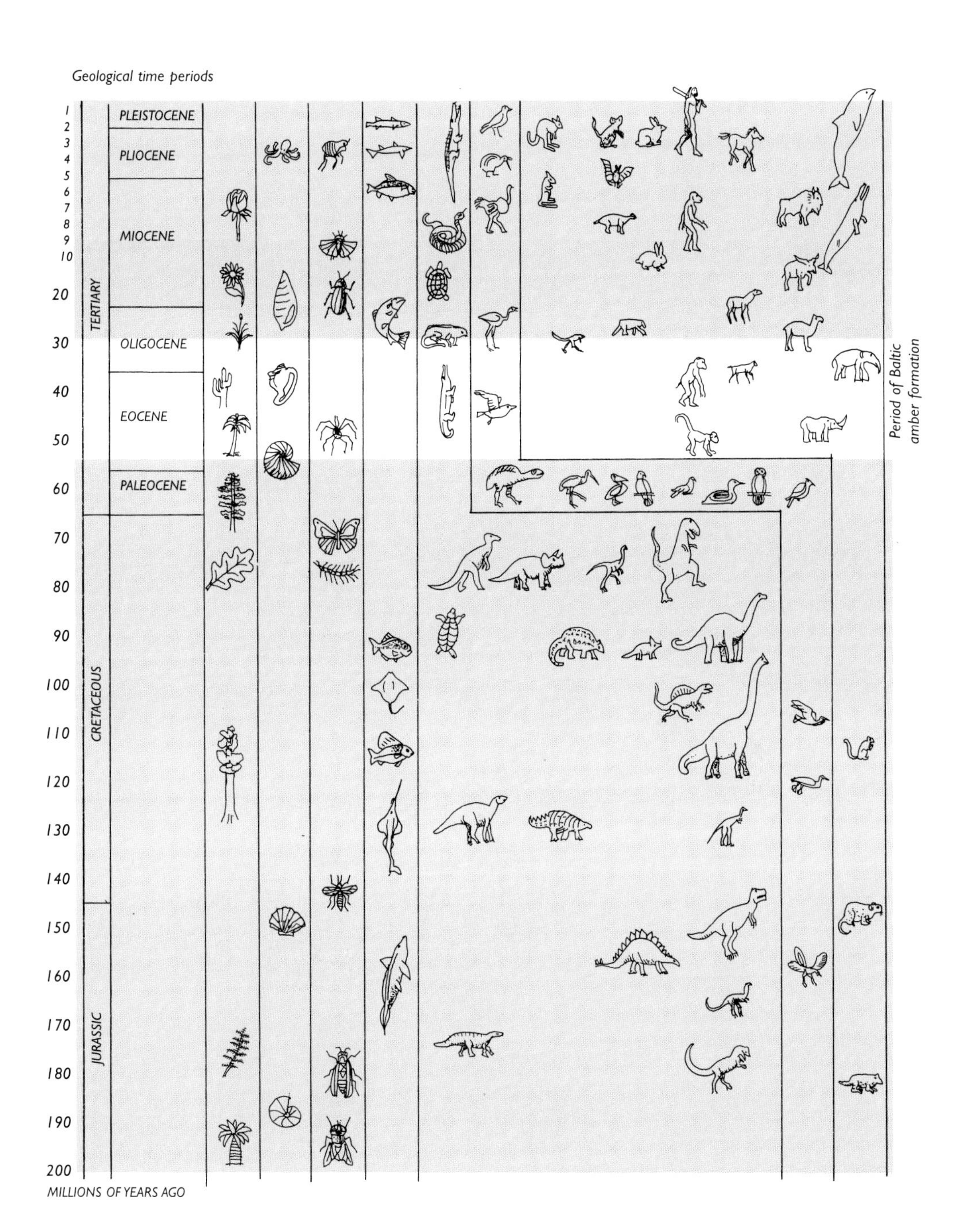
Geological time periods
PLEISTOCENE
PLIOCENE
MIOCENE
OLIGOCENE
EOCENE
PALEOCENE
TERTIARY
CRETACEOUS
JURASSIC
1
2
3
4
5
6
7
8
9
10
20
30
40
50
60
70
80
90
100
110
120
130
140
150
160
170
180
190
200
MILLIONS OF YEARS AGO
Period of Baltic amber formation

The amber at the bottom of the Baltic Sea is occasionally loosened by currents and storms and carried away by the waves. Along the way, algae and small ocean creatures attach themselves to the amber. Collecting amber on the beach has been both a profession and a recreational activity for many people; it is still the gathering method one associates with amber.

rivers. Most of it was returned to the Baltic, but some of it ended up in the North Sea via the Elbe River and other waterways. The parallel grooves sometimes found on larger amber pieces may be a result of these ice movements.

Amber may also have been deposited into the Kattegatt Strait, and onto what is today Skåne, Sweden, and Denmark, by the ice masses, or by ancient rivers now long gone. Most scientists and geologists find it most plausible that succinite amber originated from ancient forests in Scandinavia; others hold that most of it comes from forests in southern Poland and parts of western Russia.

According to recent theories, amber was carried to Samland and the Gdansk area during the Paleocene-Eocene epochs by a huge river that drained Scandinavia, Finland, and the land now beneath the Baltic Sea. The river, which Kosmowska-Ceranowicz named Eridanus (see pages 44–45), flowed into a delta between Clapowa, in Poland west of Gdansk, and an area just north of the Samland peninsula. During the same period a sound, tens of miles wide, connected the ancient Tethys Sea (which gave rise to the Mediterranean, Black, Caspian, and Aral Seas) with the Atlantic. Monsoon winds gave rise to rainstorms that carried the light amber to Germany, Denmark, and the North Sea during the summers, and to Belorussia and Ukraine during the winters.

Most beach amber has been found along Poland's and the Baltic countries' coastlines, although much has been found along the German, Danish, and south Swedish coasts as well, particularly around Kämpinge's cove, the most amber-rich area in Sweden.

Baltic amber comes in a wide range of colors. Yellow and brown are the most common.
The colors are always light when the surfaces are cut and polished.

Color and form

Amber is not composed of crystals, as are most precious stones. It is instead an amorphous material, like glass. Amber may be transparent or opaque (cloudy or muddy), and the color may vary from clear to honey yellow, orange, brown, or red. White and black amber also occur. In exceptional cases amber that fluoresces in blue, green, and violet may be found. Yellow and brown are the most common colors; red develops over time through an oxidation process that begins at the surface. If amber undergoing this process is cut in half, a red surface with a yellow center may be seen.

Taken together, there are around 250 color variations of amber. So the Swedish historian and geographer Olaus Magnus (1490–1557) was probably right when he wrote in the sixteenth century that

a painter can hardly portray objects of such shifting color as the ones nature has bestowed on amber, as they appear, after being crafted by a skilled artist's hand.

Amber's colors may appear muddied, but they become clearer when polished. A dull appearance may also be the result of many small air bubbles contained within the stone. Completely white amber may contain as many as 900,000 minuscule air bubbles per cubic millimeter, while clear amber contains only a few larger bubbles in a corresponding area. The white amber is less brittle than the clear, and is therefore easier to sculpt and shape.

The air bubbles may be removed by placing the amber in oil and gradually heating the oil to boiling temperature. Rapeseed oil used to be considered the best, but today other vegetable oils and the dry distillate amber oil are used. The oil fills the bubbles, and because it has the same refraction index as amber, the stone appears completely clear. The latest trend is to treat the stones in an autoclave by adding nitrogen.

Sculpture by coauthor Leif Brost in white amber with parts of the surface area left untouched

Amber that is weathered for an extended period of time, as most pieces found in nature are, develops a corroded surface layer. But underneath this scruffy surface, luster and color are hidden. It is the task of the amber craftsman, through cutting and polishing, to bring out the inner beauty of the amber stone. Sometimes parts of the surface layer may be left as an artistic framework.

Black amber, colored by particles of earth and plant material, is considered less valuable. It should not be confused with the "mourning stones" worn by mourners in Europe, particularly during the nineteenth century. These stones were called "black amber" but were in fact composed of jet, a type of hardened coal. The popularity of white amber, sometimes called "bone amber," has varied throughout history. The Romans held it in low regard and found it useful only as medicine and incense. But the Europeans of the Middle Ages valued it greatly and attributed wondrous powers to it.

Despite amber's range of colors, its name is used to describe a particular color tone. When we say something is "amber-colored," we are referring to the warm, brown-tinted yellow found in translucent amber. Some of the magic of amber is lent to the name, creating an association with something warm, ethereal, and romantic. Fashion designers and commercial entrepreneurs have been quick to take advantage of this. But "amber"

Multicolored stones are unusual, and highly valued.

is far from a modern word. Nero compared his queen's hair to amber. This was surely a generous compliment, considering the high value the Romans placed on the stone. It was also a signal to Roman women, who began to dye their hair to harmonize with their amber jewelry.

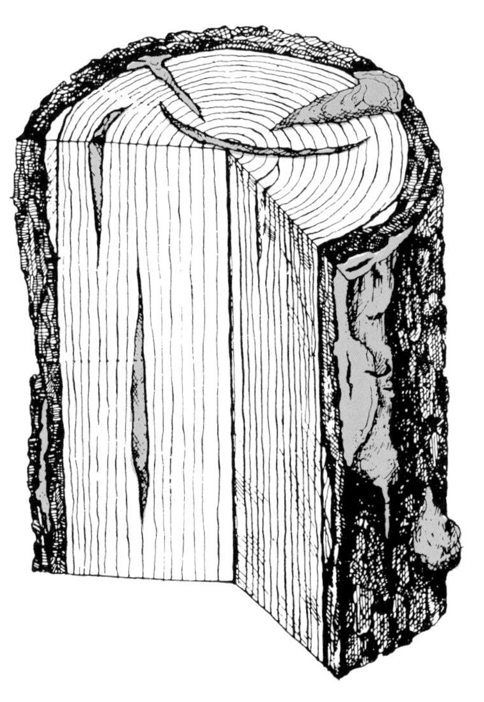

Raw amber also varies in shape and form, partly depending on whether it is formed inside or outside the tree trunk. It may be found as rounded pieces of varying sizes, as rods (formed from the resin streaks on the tree trunk), as drops, or as disks (formed in the narrow cracks inside the tree or when the resin is flattened as it falls in drops to the ground). Some stones have noticeable layers inside them, where the resin has hardened between one period of dripping and another. In some rare cases the first flow of resin will harden and create an amber embryo for a second, larger flow of resin to encase. Nature has then treated us to what the specialists call "amber in amber."

It is common to find shards and fragments of amber, but amber pieces the size of heads, weighing over 20 pounds (10 kg), have been found as well. Sweden's largest piece weighed 23.1 pounds (10.5 kg) but was broken into two parts, the largest weighing 19.6 pounds (8.9 kg), on display in Ravhuset (Amber House) in Copenhagen, Denmark. Such large pieces usually have a homogeneous structure, presumably the result of a large amount of resin accumulating during a short time span, following some major damage to the tree.

The world's largest piece is not composed of succinite. It was found on the island of Borneo and weighed around 150 pounds (68 kg). But it was also broken, this time into four pieces (see pages 56–57).

Resin that has run down the tree trunk is often transparent.

The world's largest known piece of Baltic amber (succinite) is located in the Museum für Naturkunde in Berlin (also known as the Humboldt Museum). It was found near Stettin in Poland in 1860 and weighs 21.5 pounds (9.8 kg).

Sticks and straws, which once nestled themselves into the resin, have left holes in what is today amber.

Chemically, amber is a natural polymer made from various types of resin (plastics are artificial polymers). Amber's basic elements are carbon, hydrogen, and oxygen. Sulfur is also included in the mix and is responsible for the stone's color; in addition, various organic impurities are often part of its makeup.

Amber is a light material, with a specific gravity of from 1.050 to 1.096 (1.08 is typical for Baltic amber). It can therefore float in salt water (seawater has a specific gravity of about 1.03), if not like a cork, then at least like a bar of soap. White amber, because of its many air bubbles, can float in ordinary water, while the other forms demand strong salt concentrations to stay afloat. As a result amber has often been used industrially. Indeed, as early as the seventeenth century, amber was used to measure the salt content of a solution.

Because of its light weight, amber is easily transported by the sea, and once set free from a sea deposit, whether by current or storm, it does not settle until it has reached a shore or a calm bay.

Another characteristic of amber (which it shares with other precious and semiprecious stones such as diamond, topaz, and tourmaline) is that it becomes negatively charged when it is rubbed vigorously with a piece of cloth. It then attracts lightweight particles such as bits of paper and splinters. This sign of an electric charge has been observed ever since antiquity.

Mohs' hardness scale	
1	talc
2	gypsum
3	calcite
4	fluorite
5	apatite
6	orthoclase
7	quartz
8	topaz
9	corundum
10	diamond

Precious stones most often belong to the hardest categories of minerals, but this is not the case for amber. To describe hardness, we use Mohs' scale, which ranks materials from 1, representing the hardness of talc, to 10, representing the hardness of diamond. Amber is ranked at around 2.5, somewhere between gypsum, which can be scratched with a fingernail, and calcite, which can be scratched with an iron spike. The surface is somewhat harder than the core, which indicates that the hardening process occurs from the outside. This process is continuously at work, so generally, the harder the amber, the older it is. But amber is easy to soften. It becomes rubbery at 285°F (140°C), and melts at from 280° to 720°F (250° to 380°C). When burned, it emits the smell of incense. Myrrh, familiar to us from the Bible, may be thought of as a relative of amber; it is composed of the hardened resin from an African tree

Windows to a past world

As mentioned earlier, insects and other small animals that have become trapped in resin and perished can sometimes be seen encased in transparent amber. This has preserved them for posterity in a way whose completeness is unparalleled, save perhaps for the inclusions frozen in the permafrost. The amber has thus trapped a scene from life millions of years ago. And it isn't just an imprint of some sort, it is the physical object itself that has been included—a frozen piece of time!

In some stones, entire swarms of winged insects have been included; their struggle to break free can be seen in the solidified patterns and torn-off legs that remain in the amber. Sometimes the legs are the only part of the insect that are left; the body didn't sink into the resin and became food for birds, or was destroyed in other ways. It may also be the case that the insect has sacrificed one or more of its legs and flown away crippled.

The British poet Alexander Pope (1688–1744) expressed his surprised admiration for amber and its inclusions in the following manner:

Pretty! in amber to observe the forms of hairs, or straws, or dirt, or grubs, or worms! These things, we know, are neither rich nor rare, but wonder how the devil they got there.

We know how they got there, just as we know how everything else—animals, plant parts, earth, and water—ended up included in amber resin. In the case of the fly, which elevates the amber to precious jewelry or museum status, its entrapment was obviously an accident. To paleontologists and other scientists, however, these mishaps have brought much excitement. The combination of the resin's ability to seal the inclusion, which protects it from oxygen and other eroding forces, and its transparency, which allows the inclusion to be observed from all directions, provides researchers with a unique opportunity to acquaint themselves with the life of 20–320 million years ago.

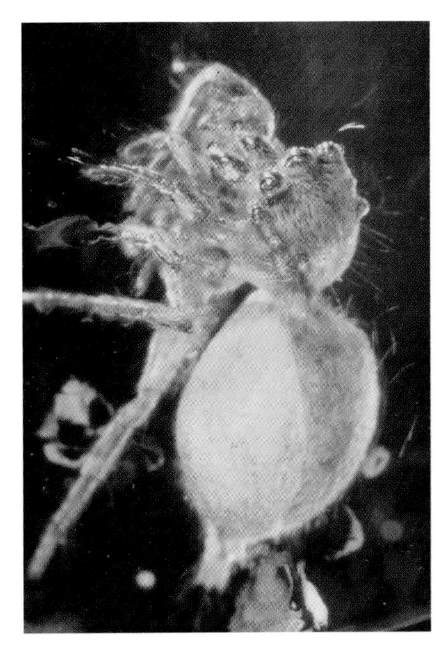

These insects may have landed in the resin by accident, but much indicates that they were attracted by the resin's aroma, or by the glimmering light it cast, and were lured in like flies to flypaper. In some stones entire swarms of winged insects may be found.

Body of a grasshopper that landed in the wrong place

Linnaeus, the father of modern botany, must have been fascinated by amber's ability to include and, over millions of years, preserve insects and other creatures. It gave him the fantastic idea—if it indeed was his own idea —of "storing the bodies of kings and monarchs in amber, to preserve them for posterity." All that was needed was to "find how the stone could be melted and remolded around a man's body," he wrote. According to certain unconfirmed sources, Lenin, or at least his face, was protected in his coffin by a layer of molded translucent amber.

The scope of this insight, however, is limited to animals that were too small to free themselves from the resin and to the animals and plants that existed in close proximity to the trees in question. Still, thanks to amber, we are able to get precise information about something that millions of years have otherwise erased.

That the resin itself has stood up to time so well is partly a result of its intrinsic preserving properties, and partly a result of being stored in the earth or water—thus avoiding the ravages of oxidation. Baltic amber is the only appreciable remnant of the vast forests thought to have covered northern Europe millions of years ago. It is a condensed version of a world we can otherwise only imagine. As we hold a piece of amber in our hand, this fact is well worth pondering.

It is interesting to note that inclusions are almost always found in transparent amber. Amber craftsmen say that they almost never encounter an inclusion while sculpting an opaque stone.

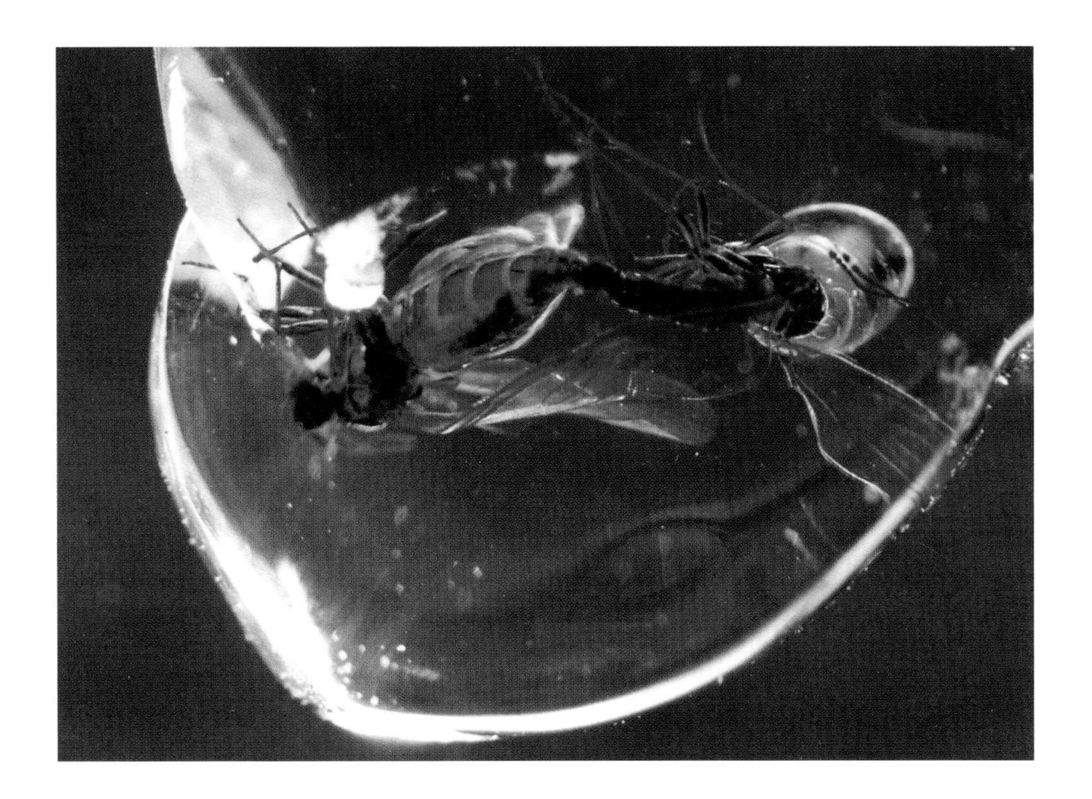

Caught and immortalized together

A new theory concerning the size of some amber pieces, and the amount of amber insect inclusions, is that they are a result of a defense mechanism by which the trees in question produced and released large amounts of resin when threatened. The theory may not be so far-fetched. It rests on observations that certain contemporary trees react similarly and certain contemporary plants practice various forms of chemical warfare with their competitors in the battle for living space.

There have been conflicting opinions about how much of the entrapped insects really remains. Many have maintained that most inclusions are merely imprints or molds with preserved color pigments, and that the rest of the animal has become decomposed or carbonized. But today we know that there are many cases where more than the exterior parts are preserved. With specialized instruments we have been able to see, for example, an insect's muscles, and its

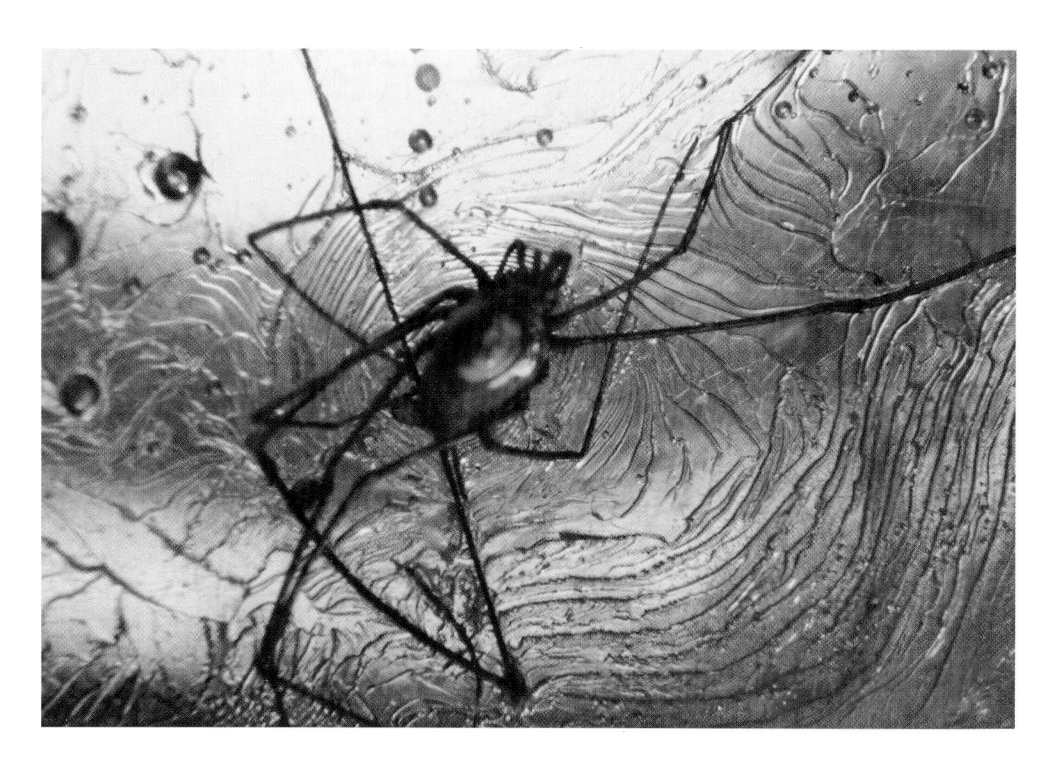

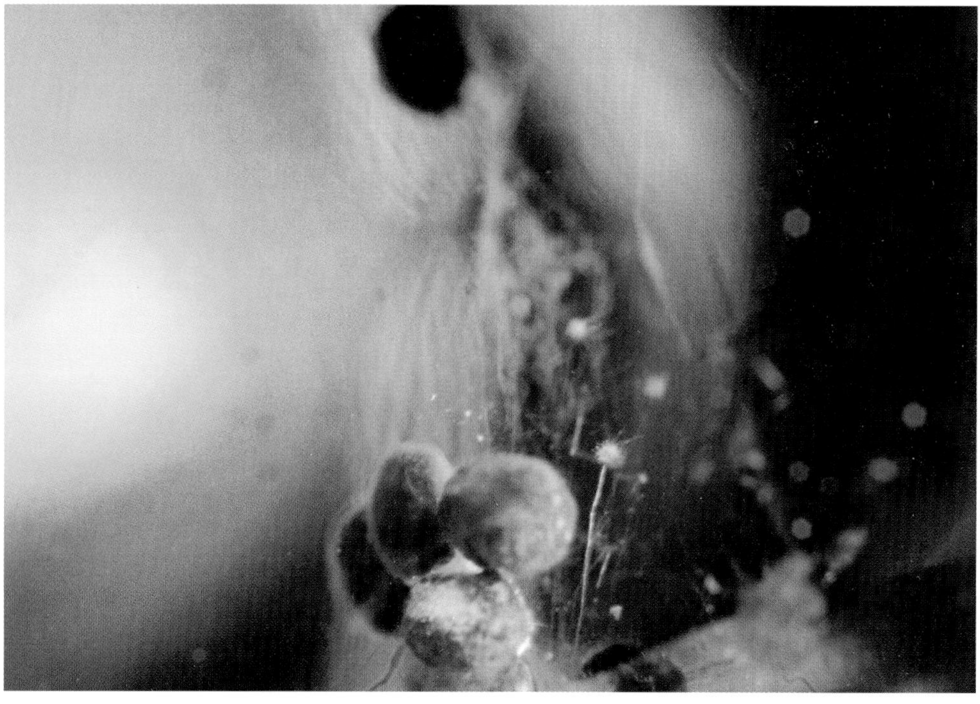

organs such as heart and liver. However, extracting an insect inclusion requires extreme caution. It is often the case that such attempts result in nothing more than a tiny pile of dust.

The interiors and skeletons of animals such as frogs and reptiles that, unlike insects, have their skeletons on the inside, may be harmlessly observed with X-rays. With electron microscopes scientists have been able to distinguish individual cells and their components (nucleus, mitochondria, ribosomes, and so on). Some American researchers have even been able to isolate parts of the cell nucleus's DNA (the code of all living beings). More about this in the next chapter.

The accumulated treasures of amber from around the world reflect life in the Triassic period 225 million years ago (when Austrian and Bavarian amber was formed, as well the Chinle Formation amber in Arizona); the Jurassic and Cretaceous periods, which ended 65 million years ago (when Canadian and Siberian amber was formed); and the Eocene and Oligocene epochs of the Tertiary period 30–50 million years ago (when most Baltic amber was formed). If we include the presence of the much younger *copal* or "African amber" (see pages 109–110), we have, paleontologically speaking, relatively modern objects of comparison. Unfortunately, this does not grant us a comprehensive chronological picture, where the evolution of life-forms can be followed in linear progression. The findings are limited to certain parts of the world, whereas living conditions may have varied from place to place. All the same, amber provides a gateway of understanding for the scientific community, which has only begun to utilize all the existing material.

What kinds of animals tend to get stuck in resin? Primarily we find insects such as flies, gnats, midges, ants, and (less frequently) mosquitoes. They are responsible for around 80% of the total animal inclusions. We also find wasps, termites, cockroaches, cicadas, biting lice, fleas, silverfish, mayflies, moths, grasshoppers, many other types of flies, beetles of all kinds, bees (sometimes carry-

A daddy longlegs saw an easy prey caught in the resin but itself ended up falling victim to the resin.

Amber interferes with the reflection of light. For this reason, the true colors of amber inclusions are seen less vividly or sometimes not at all. If the insect is successfully extracted from the stone, however, its original color is most often still intact. Color conservation is thus another of amber's many attributes.

Tiny organisms are sometimes included in amber and may be studied under a microscope. In the photo we see insect excrement (most probably from a beetle or termite) with the mycelium threads and fruiting body of an Aspergillus *fungus.*

ing pollen), and mites (sometimes still attached to their hosts). Less common are sucking lice, fleas, dragonflies, butterflies, praying mantises, and stick insects, as well as small animals such as spiders (sometimes alongside webs and cocoons), centipedes, scorpions, wood lice, snails with their shells, worms, nematodes, and ticks. In Mexico even a small crab has been found in amber. All stages of these arthropod lives are represented. In addition to fully formed specimens, we have found eggs, larvae, and pupae. In some cases we have even observed that certain animals are hosting fungal diseases, parasites, and such.

Among the vertebrates, we find a few entire inclusions of frogs and small reptiles. Beyond that, the occurrence of vertebrates is revealed mainly by traces and parts. Examples include bird feathers and down, hairs from mammals, skins from snakes and reptiles, wood splinters with tooth marks from rodents, parasites particular to higher-order species, footprints, and feces. A piece of amber containing jaw fragments and teeth that most likely belonged to some type of swine has been found, as has one containing a tail most likely once belonging to some type of rodent. But amber inclusions of larger animals such as fish (one exists at the Royal Academy of Science in Uppsala, Sweden) are almost certainly fakes.

A characteristic of Baltic amber is that it often contains the tiny stellate hairs from the buds of oak trees.

To the layman, the entrapped species appear very similar to those of today. We are struck by how long these species have lived in the world in comparison to us humans. In truth, we are the ones imposing ourselves on their living space, not the other way around, even though we don't usually like to think so. We are also amazed at the remarkably gradual evolutionary changes within these animal groups, especially when we compare them to mammals, particularly humans, that have undergone such rapid change in such short time spans.

Of special interest to researchers and jewelry designers alike are amber pieces that contain scenes from life, such as a pair of flies caught in the act of copulation, insects about to be wound into a spider's web, ants carrying away a centipede, and parasites leaving their imprisoned host like rats from a sinking ship. And once again take note, each of these is not merely an imprint, but the event itself! It is preserved in the present, not the past, tense.

The plant kingdom is represented in amber by fungi (most often the mycelium, but a beautiful inclusion of a fully developed mushroom has recently been found in East Brunswick, New Jersey), algae, moss, and higher plant species. Of the latter, it is mainly pieces that have been encased, such as branches, leaves, twigs, pieces of bark, flowers, fruits, and pollen. By categorizing these species we can determine what the ancient amber forests looked like. The range of species is striking. For example, in Baltic amber we have documented 750 distinct plant species. Because we know their demands regarding temperature, humidity, and so on, we are able to understand what the climate of the era was like.

Besides inclusions, we also find exterior imprints, primarily of plant parts. They have been formed by the resin dripping down on, for example, a leaf on the ground. The leaf later decomposed, leaving only the pattern behind. Other times leaves have blown into the hardening resin. Clear examples of such imprints have been found of both elm and palm leaves. Footprints of small mammals have also been found, as have tiny craters caused by raindrops.

An imprint from a piece of a palm leaf of the Arecaceae family. In the subtropic forests of northern Europe, the amber pine (Pinus succinifera) *grew alongside palms, cypresses, dragon trees, and horsetails*

SPECIES FOUND IN AMBER

Pinus	*pine*
Larex	*larch*
Sequoia	*sequoia*
Chamaecyparis	*cypress*
Acer	*maple*
Ilex	*holly*
Clethra	*lily-of-the-valley tree*
Castanea	*chestnut*
Quercus	*oak*
Cinnamomum	*cinnamon and camphor trees*
Magnolilepis	*magnolia*
Oxaliditis	*sorrel*
Rhamnus	*alder buckthorn*
Picea	*spruce*
Abies	*fir*
Sciadopitys	*umbrella pine*
Juniperus	*juniper*
Phoenix	*date palm*
Sambucus	*elder*
Andromeda	*heath, rhodendron*
Fagus	*beech*
Geranium	*geranium*
Linum	*flax*
Myrica	*sweet gale, bog myrtle*
Polygonum	*sorrels*
Deutzia	*hydrangeas*

Baltic amber trees include some species that reside exclusively in tropical climates, and others that reside exclusively in subtropical climates. Some scientists therefore conclude that the climate in the Baltic area was similar to that of today's Florida, where plants from various temperate zones coexist. Others believe that the temperature varied according to time and geographical region.

Even if the species, in some respects, are different from the ones of today, they can be clearly defined and sorted into the existing families and orders. On the left are some examples that should be familiar to garden enthusiasts.

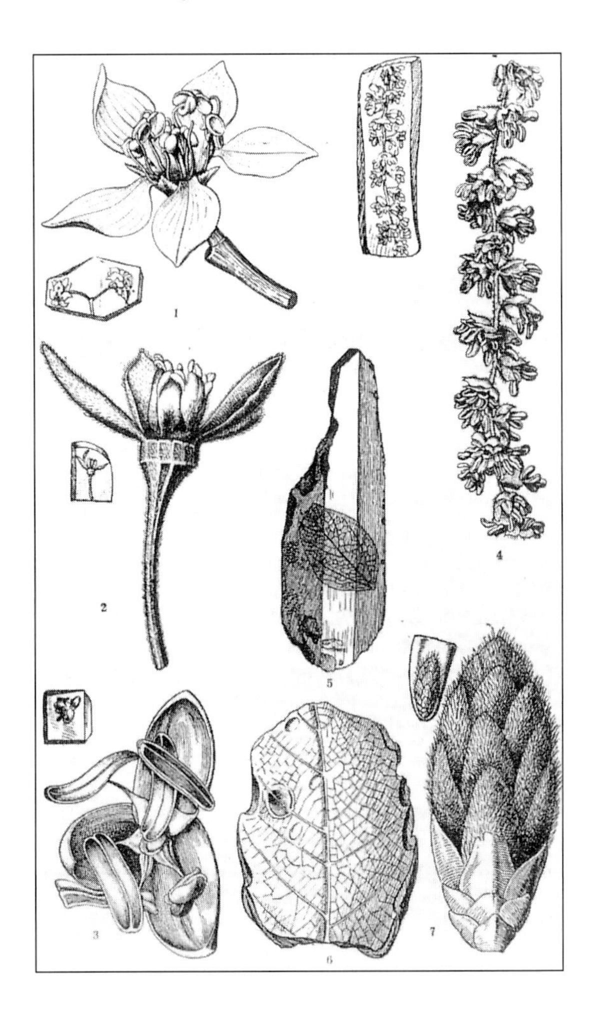

During the nineteenth century, careful descriptions were made of the amber forest's fauna. They are of great assistance to scientists today.

Air pockets create plantlike patterns.

Plants that are molded for insect pollination must logically have evolved in the same period as the pollinating insects. The recent discovery of a few flowers in 90-million-year-old amber is therefore very significant to evolutionary history. It supplies a foundation for comparisons and may provide important clues to how the symbiotic relationship between pollinating insects and flowers began.

You might think that, through clues provided by plant inclusions and analysis of amber material, determining what species the amber resin originated from would be simple. As we have mentioned, however, this is not the case; as usual, the experts disagree. The only true consensus concerns Mexican and Dominican amber; the experts agree that it originates from a deciduous tree called *Hymenaea,* a member of the pea family (Leguminosae). Parts of *Hymenaea,* an extinct tree that has living relatives in both America and Africa, have been found in amber inclusions. Researchers have also been able to extract DNA from its cells.

Bacteria have also been found in amber. By partially dissolving amber in turpentine, and then straining and centrifuging the resulting substance, these bacteria can be studied under a microscope. In this manner, researchers have located bacilli, cocci, and spirilla bacteria of various kinds. (Evidently none of these bacteria could eat amber.) In addition, amoebas and other protozoa have been found. Certain amoebas have even been preserved while in the act of dividing.

Inorganic material such as earth, gravel, volcanic ash, water, and air has also been trapped in amber. Sometimes we find a water inclusion with an air bubble inside it—nature's own spirit level. The water inclusions also enable researchers to analyze liquids millions of years old. The chemical structure, and the rate of impurities in this water, may provide additional circumstantial evidence regarding the evolution of the earth. Experiments have been conducted, but one must be cautious when drawing conclusions. The air inside the water inclusions is not always the original. Experimentation has revealed that a certain amount of exchange with the air outside may occur.

All in all, amber provides us with genuine and valuable information about the plant and animal worlds of, primarily, the Cretaceous and Tertiary periods. We gain insight into species' biogeography, evolution, and ecosystems.

The often numerous air and water bubbles can be thought of as inclusions as well.

American researchers, led by Raul J. Cano, have examined, among other things, the stomach contents of several 25- to 40-million-year-old stingless bees preserved in amber from the Dominican Republic. Through DNA analysis, the researchers discovered a certain species of bacteria living inside the bees' stomachs. These bacteria are related to the bacteria that live in the bees of today. Now, as then, this strain of bacteria exists in symbiosis with the bee. The bacteria help the bee digest pollen and other forms of nourishment, and protect it against certain diseases by producing antibiotics. The bee returns the favor by supplying food and a home for the bacteria.

The discovery is interesting on its own. But the real sensation came when Cano, together with Monica Boruche, proved that there was still life in these 25- to 40-million-year-old bacteria. When placed in a nutrient solution, the bacteria reproduced themselves quickly. It is known that bacteria, through encasement in endospores, can survive for a long time in a state of inertia without being affected by their environment, but 25–40-million years is longer than anyone had dared dream. Some researchers believe it is still a dream, that, despite all the precautions taken, a modern strain of bacteria must have entered the experiment and been responsible for the resulting reproduction. The experiments and the disputes will most likely continue.

The question still arises: If in fact the old bacteria are reproducing themselves, is there a danger involved? Could they break out and spread new diseases among animals and humans? The researchers don't think so. They maintain that their experiments will assist in the process of finding new forms of medicine. But researchers aren't always right.

With its fossils, amber has made an enormous contribution to science. Baltic amber alone has delivered over a million stones with inclusions. Unfortunately, some collections were destroyed during World War II, but there is still much left, and new finds are being made all the time. Copenhagen's zoological museum has a fine collection of inclusions, and in Sweden such collections may be found in the Department of Paleontology at Uppsala University and in the Amber Museum in Kämpinge.

*Gecko (*Sphaerodactylus *sp.) preserved in amber from the Dominican Republic. In the older literature there are pictures of frogs and reptiles in Baltic amber. Many of these stones have not been located, and those which have, have turned out to be fakes. However, a genuine amber inclusion with a piece of reptile skin exists in a museum in Copenhagen.*

Dinosaurs in amber

Are there dinosaurs in amber? Obviously not. Even dinosaur footprints in amber are among the very rarest specimens found. But the "blueprints" needed to create and clone dinosaurs might be out there. At least so it has been speculated.

It all began a couple of years ago when some American researchers managed to isolate DNA from a few insects in amber inclusions. First on the scene was Berkeley insect pathologist George O. Poinar, who was able to extract pieces of DNA from a 30- to 40-million-year-old bee. Around the same time, a research team in New York announced that they had found, in a piece of amber of roughly the same age, genetic material belonging to an extinct species of termites. These DNA fragments could be cloned, or duplicated, and a new research discipline, molecular paleontology, was born.

These findings set the imagination in motion. What if we could find a mosquito, or any other blood-sucking animal, that shortly before being entrapped in amber resin had sucked the blood of, for example, a dinosaur? DNA could be isolated from the blood and—a hair-raising thought—be used to clone and create a live dinosaur.

Charles Pellegrino, a paleobiologist and science fiction author, put these ideas down on paper when he wrote: "We may be able to insert genetic material into a nucleus, furnish an egg yolk and an eggshell, and hatch our own dinosaur." And Michael Crichton took these ideas one step further with his best-seller *Jurassic Park*. After the novel was made into a film by director Steven Spielberg, it attained Jurassic proportions in all regards. A heightened interest in dinosaurs as well as amber has followed in the film's wake. An exhibition containing both models from the film and amber stones with insect inclusions has toured the United States (coauthor Leif Brost has contributed several unique stones to the exhibition).

In *Jurassic Park*, which can be interpreted as a warning against allowing commercial interests to take over DNA research, a businessman has engaged a team of scientists—geneticists, paleontologists, computer experts, and others—in extracting dinosaur DNA from the blood of mosquitoes that have been preserved in amber. This DNA is later complemented with the living DNA of related animals, such as frogs, and implanted in crocodile egg whites, which in turn are implanted in plastic shells. The synthetic eggs are then incubated under heat lamps and—hocus-pocus—hatched into living baby dinosaurs.

To capitalize on this venture, the businessman, with the assistance of Japanese and American investments, builds a Jurassic theme park on an island off the coast of Costa Rica. Both carnivore and herbivore dinosaurs are placed on the island. The park includes a specimen of the ferocious *Tyrannosaurus rex*, and several specimens of the quick, and viciously predatory, *Velociraptor*. Naturally, things take a turn for the worse. Despite measures taken to ensure that all the dinosaurs are female, the species reproduce and break free from their captivity—nature's ability to adapt has not been taken into consideration, and certain dinosaurs undergo a sexual transformation.

But let us return to reality for a moment, or rather a more conceivable future. In 1993 George Poinar, mentioned above, succeeded in extracting and studying a few sequences of DNA from a beetle, trapped in Lebanese amber, that dated back to the Cretaceous period. This started a debate in the scientific community as to whether such rare inclusions should be experimented with in this way. According to certain sources, Poinar now plans to extract the blood from all the parasite inclusions in his collection. He will then concentrate the DNA and compare it to that of the dinosaur's closest living relatives, birds and crocodiles. In this manner, he hopes to gain material to chart the evolution of dinosaurs, and perhaps to come up with a definitive answer to why they became extinct. And perhaps he will come up with something more!

Even if it is theoretically possible to recreate

extinct animal species, the hurdles are enormous. Many lucky coincidences and groundbreaking research results would be required.

To begin with, amber inclusions from the dinosaur periods (Jurassic and Cretaceous) are not too common. In any case, the chances of finding an insect that had just sucked the blood from a dinosaur are minimal; moreover, its DNA would be either partly or fully destroyed during the vast amounts of time that have passed since these periods. Research indicates that less than 1% remains after the passage of 25 million years. In the amber in question, only fragments remain, and it would be a stroke of luck if just those parts were the ones specific to dinosaurs. (DNA is largely similar for most organisms; it contains around 90% inactive information—perhaps a reserve for future evolution.)

Up until now, the paleo-DNA that has been extracted contains a maximum of 300 base pairs, while the DNA of a dinosaur must have contained at least a billion base pairs (human DNA contains three billion). It is also debatable whether DNA can be extracted from blood cells.

Apparently, cloning dinosaurs from extracted DNA is a dream for some people and a nightmare for others. But with the possibilities of cloning a human being what they are—DNA has, for example, been extracted from an Egyptian mummy—a dinosaur might be preferable to, say, a clone of Hitler or Stalin. Perhaps even more bizarre is the thought of cloning ourselves. Then we could achieve the ancient dream of immortality. There is always room for dreams and speculations. Amber invites them.

A stone or not a stone? —A matter of semantics

We refer to amber as a "precious stone," but is it really a stone? Geologists categorize stones as loose rock formations, and amber does not belong in that category. But when we consider its derivation—an Indo-European word meaning "to harden" or "thicken"—and its usage, in words like "kidney stone," "stone fruit," and "gallstone," it makes sense to apply the word to amber. In addition, amber has always been considered a precious, or at least a semiprecious, stone.

The German word for amber is *Bernstein*. Its prefix is derived from the Low German word *bernen*, or *bärnen*, meaning "to burn" or "shine." Earlier words for amber include *gles*, *glis*, *glys*, and *glas*, the first of which appears as a prefix in the name of the Danish district Glesborg. These words have their origins in the Germanic word *glessum* and the Latin word *glesum*, both meaning "amber." *Rav*, another word for amber meaning "brown," is part of the name of the south Swedish district Ravlunda. Swedish rosaries, used to count prayers in the old days, were once called "*ravband*" because the beads were usually made of amber. Even in the Koran, there is a passage that says that the pilgrims to Mecca are to wear rosaries made of amber.

In the seventeenth century the Swedish words *agd* and *agdsten* were used; both are believed to come from the word "agate," another gemstone used to make jewelry. An author from that period wrote in 1622: "They attracted as many hearts as an agate or victory stone (that is to say, a magnet), a piece of steel or iron." "The '*agdsten*,'" wrote another in 1684, "was collected along the sea in strange ways out in Prussia."

The Greek word for amber is *elektron* (akin to *elektor*, meaning "beaming sun"). It is also the origin of the word "electricity," the existence of which one of the seven wise Greeks, the philosopher Thales, demonstrated by rubbing a piece of amber with cloth and allowing it to attract light objects. While on the subject of the word's origin,

Precious stones

Opinions vary as to what should be considered a "precious stone." The more general definition includes all organic and inorganic stones used to make jewelry. The narrower definition includes only stones with a hardness of 8 or above on Mohs' scale, such as diamonds, emeralds, rubies, and sapphires. Other stones are referred to as "semiprecious stones." Today, the commonly accepted sense of "precious stones" is the more general one. To avoid any controversy, it may be simpler to use the term gemstones. *Many precious stones are of non organic origin. Only amber, pearl, coral, ivory, jet, and tortoiseshell have organic origins.*

it may be interesting to note that the tip on Pallas Athena's lance was made from an *elektron*. Myth, magic, and science are joined in this single word that still exists today.

Another Greek word for amber was *harpaks*, which means roughly "miser" or "someone who pulls things toward him" (compare to the Danish word *harpiks*, meaning "resin" or "rosin"). The Latin word for amber was *succinum*, from which comes *succinite*, the term used to describe Baltic *amber*. The French word for amber is ambre, whose origin is the same as the English word and which can also refer to the aromatic gallstones that the sperm whale produces. Most likely it is the pleasing aroma, and the fact that both "stones" are found in the sea, that has caused the two objects to have the same name. The word *ambra* has even been used in Swedish to mean "amber," but only for certain classifications of yellow amber.

Ambrosia, the drink of immortality consumed by the Greek gods, is also related to the word *amber*; and both words are thought to be related to the Greek word *ambrotos*, which literally means "immortality." Many amber finds in the form of amulets indicate that the stone was thought of as more than just decoration. Perhaps it was worn in hope of gaining immortality, as a ticket to life in the kingdom of the dead. This kingdom, which at times was called "Amber Mountain," has been altered in the myths and is now called "Glass Mountain" (according to some sources, the Germanic word *gles*, meaning "amber," is derived from the Middle English word *glas*).

The Russian word for amber is *yantar*, which may be of use to those who wish to go there and sample their amber. The word reoccurs in "Yantarnyi," which is the name the Russians gave to the center of the amber extraction in Samland, and in the "Yantardak," the name of a summit in Siberia, which literally means "Amber Mountain."

Gold, honey, light, and tears

Amber has also been given more poetic names that stir the imagination: "sea gold," "petrified light," "sun tears," and "tiger's soul" are a few such names. The last name comes to us from China, where amber was a symbol of courage. Other embellished names are "Nordic gold," "captured sunshine," "hardened honey," and "Freja's tears." According to ancient Nordic myth, having betrayed her husband, the weeping and regretful goddess Freja went in search of him. Where her tears touched the earth, they turned to gold. But where they touched water, they turned to amber.

The tear motif is also found in the ancient Greek story of Phaeton. After he died his sorrowing sisters, the Heliades, were changed into trees that cried amber tears. As is often the case with Greek mythology, many versions of the same story exist. This is the most common one:
Phaeton was the son of the sun god Helios and the Egyptian queen Clymene, with whom Helios also had seven daughters called the Heliades. When some of the other children of the gods questioned Phaeton's divine origins he wanted to prove that he was indeed the son of the sun god. To do this, he decided to ride the sun, which was drawn by nine white fiery horses. By choosing his words well, he was able to make Helios promise to let him have a hand at the reins, something the sun god had never promised anyone else. But as it turned out, Phaeton lacked his father's strength and steady hand. He lost control of the reins and the horses galloped so far up into the sky that the earth froze. Then they galloped so low that the earth burned, the Sahara was formed, and the Ethiopians gained their dark complexion. The whole affair would have ended in catastrophe if Zeus hadn't struck Phaeton dead with a bolt of lightning and sent his body plunging into the river Eridanus. Phaeton's sisters, the Heliades, came to cry at his grave, and so afflicted were they that they turned into poplars, aspens, or (according to some accounts) alders. Their tears turned into amber and were collected along the river's banks.

Poetry has to some degree found inspiration in the uncertainty over what amber actually is. Aristotle correctly assumed that amber came from trees, and Pliny the Elder, who lived in the first century A.D., was convinced of this fact. Tacitus, who lived sometime between 55 and 120 A.D., was convinced of this, too, and developed his thoughts on the subject in his writings *Germania* and *Salinus*. But the the sixteenth-century German mineralogist Agricola dismissed these theories by stating: "Amber is found in the sea, and no trees grow in the sea." Many theories concerning amber's origin were put forth in the ensuing centuries by scientists who were uncertain and, naturally, in disagreement. Not even Linnaeus could provide an answer. Some believed that amber was wax from forest ants, others that it was gallstones or lynx urine. Still others thought amber was honey, affected by the earth's life force, or hardened and frozen sap from cherry trees. A common theory was that amber was condensed sea foam. During the eighteenth and nineteenth centuries many believed that amber was a mineral, or a type of solid petroleum. They suspected that a giant oil well was located in the middle of the Baltic Sea, and that the oil, when it rose to the surface, could catch insects and small animals; this explained the inclusions. It was not until 1811 that a scientist in Danzig confirmed the truth about the origins of amber.

The rather curious belief that amber came from lynx urine may be the result of a linguistic misinterpretation. *Lynx* is a Latin word, and amber used to be called *lyncurium*, most likely after *lucere* meaning "to shine," but felt by some to derive from Liguria, the name of a region in Italy where amber was found or perhaps only traded. *Lyncurium* was read, or interpreted, in Swedish as *lourinsten* (which sounds like the Swedish word for "lynx stone"), according to certain researchers. Another possibility is a confusion of amber with *lapis lyncis*, the fossilized remains of the belemnite, an octopus of the Cretaceous period, which often possess an amber yellow color. These fossils are composed of bituminous limestone and, when

crushed, exude a smell of ammonia, reminiscent of cat urine.

It is surprising that so many fallacious theories about amber developed during the seventeenth and eighteenth centuries. A relatively accurate theory had been developed as early as the sixteenth century. Olaus Magnus, the priest and humanist sent to Rome by the Swedish king Gustavus Vasa, articulately describes the origin of amber. In his 1555 publication *Historia de gentibus septentrionalibus* (The History of the Nordic People), he gives the following description:

The firs or conifers, which by their nature contain much resin, whether they rise above the coast of the sea, the edge of the river, or the slope of the forest, secrete an amber substance. . . . This is particularly the case when, at the time when the fruit ripens, the sun, more scorching than the summer heat during the months of June and July, moves into the signs of Cancer and Leo. In this heat the bark on the trees breaks, and thereafter drop after drop of amber substance flows out and down into the rills and rivers, where it eventually hardens. Sticky and viscous as glue, this substance traps everything that comes in its way. Pulled into this mass [are] frogs, mice, mosquitoes, spiders, flies, bait, fruit, and other things that eventually harden inside it. This occurs even if the drops fall into the sand, because when the rain comes and washes over it, the resin is flushed into the stream, it is then carried to the river, and eventually out into the sea. In the sea it hardens, like wood that floats in the water for a long time, and after a

The river Eridanus is also referred to as the "Amber River" in other contexts. But it cannot be located on any map, and there have been many discussions as to which river it was. Perhaps it was the Po River, or maybe the Elbe, or some other river that flows into an amber-rich body of water. If we travel south and look up, we can view Eridanus winding its way through the night sky, like a heavenly slalom hill with ports of gleaming stars. Or is it amber tears that are shining like stars? The imaginative seventeenth-century reverend Jonas Floraeus from Ravlunda, in southern Sweden, was convinced it was in Ravlunda that Phaeton met his fate. According to Floraeus, even the name Ravlunda, which literally means "amber grove," suggested that.

Tacitus wrote the following in his work Germania*:*

The Aisti people even search the sea and collect, as no other people, amber—which they call "glesum"*—on the beach and in shallow waters. They, as one may expect from barbarians, have not devoted any research or thought to where the amber comes from, and how it is created. For a long time they left it lying along the beaches, together with the other objects that the sea washes up, until finally our desire for the stone became known. They have no use for it themselves. They find it as it is, hand it over uncrafted, and listen in dumbfounded amazement to the price that is offered. We, however, know that amber is made from tree sap because earth insects, and even winged insects, are often seen inside it. These creatures became stuck in the sap when it was still in liquid form, and were trapped inside it when it hardened. I therefore presume that lush groves and forests grow on the islands and mainlands in the West, and as the sun shines down, the sap from the trees is brought forth. While in liquid form it runs down into the nearby sea and is washed, with the violent storms, to the opposite shores.*
(From a Swedish translation by Bengt Ellenberger)

certain amount of time it turns into the hardest stone. When a violent storm later occurs in the Geatish, the Finnish, or the Sea of Livland, the amber is thrown up on the Prussian coast that lies south of Scandinavia. This occurs constantly, as the waves drift with the wind to the land of Prussia. On this coast, by request of the prince, only specifically chosen workers, who have sworn an oath of loyalty, are allowed to collect the amber.

The main lines of reasoning behind this evolution story are correct, and the descriptions are rather inspired. That Olaus Magnus, as many scientists of recent centuries, was mistaken about the age of amber is understandable. According to the Bible—and the Bible could not be questioned—the creation of the earth wasn't too far back, perhaps only a couple of thousand, or tens of thousands of years old. But he could have researched his information about the inclusions more carefully: mice have never been found in amber.

The comparison to gold, which recurs in many embellished writings, has its roots in the color, shimmer, softness, and rarity of both materials. But the differences are considerable: amber is very light, while gold is one of the heaviest metals; amber is a good electric insulator (and has been used for that purpose), while gold conducts electricity better than most metals.

Woodcut portraying amber collectors on the coast of Prussia. From Olaus Magnus, Carta marina*, 1539.*

In his Carta marina*, published in Venice, Magnus indicated the amber-rich areas around the Baltic Sea with little barrels, which was how they used to store amber in those days.*

Amber land and amber sea

The land of amber exists on the southern coasts of the Baltic Sea and on Germany's and Denmark's North Sea coasts. The amber is hidden in the sea. As mentioned, the largest amber reserves are located in the Baltic countries, but much also exists along the shores of Poland, Germany, Denmark, and southern Sweden. The greatest supply of amber can be found in Samland and the surrounding areas. Sweden's reserves are located on its southeastern coast in the region of Skåne, while Denmark's western Jutland harbors the "Ravkyst" or "Amber Coast."

Amber has also been found in several lakes and rivers, at times in large quantities. Examples include many German rivers and the Mazovian lakes in Poland.

Skåne's amber is found not only along the coast, but in ample amounts farther inland. These deposits are found in a fault in the earth's bedrock, the so-called Alnarps River, a preglacial riverbed 3 miles (5 km) wide, covered by later geological strata. The Alnarps begins just west of the city of Ystad, and extends northwest to Glumslöv near Helsingborg, continuing under the Öresund, the sound between Sweden and Denmark, and on to northern Själland (see map). The finds in this area were of such scope that the Swedish state geologist N. O. Holst, who had studied the Alnarps River and its origin in detail, wished to call it the "Amber River."

Amber exists in large quantities along many coasts of the Baltic Sea.

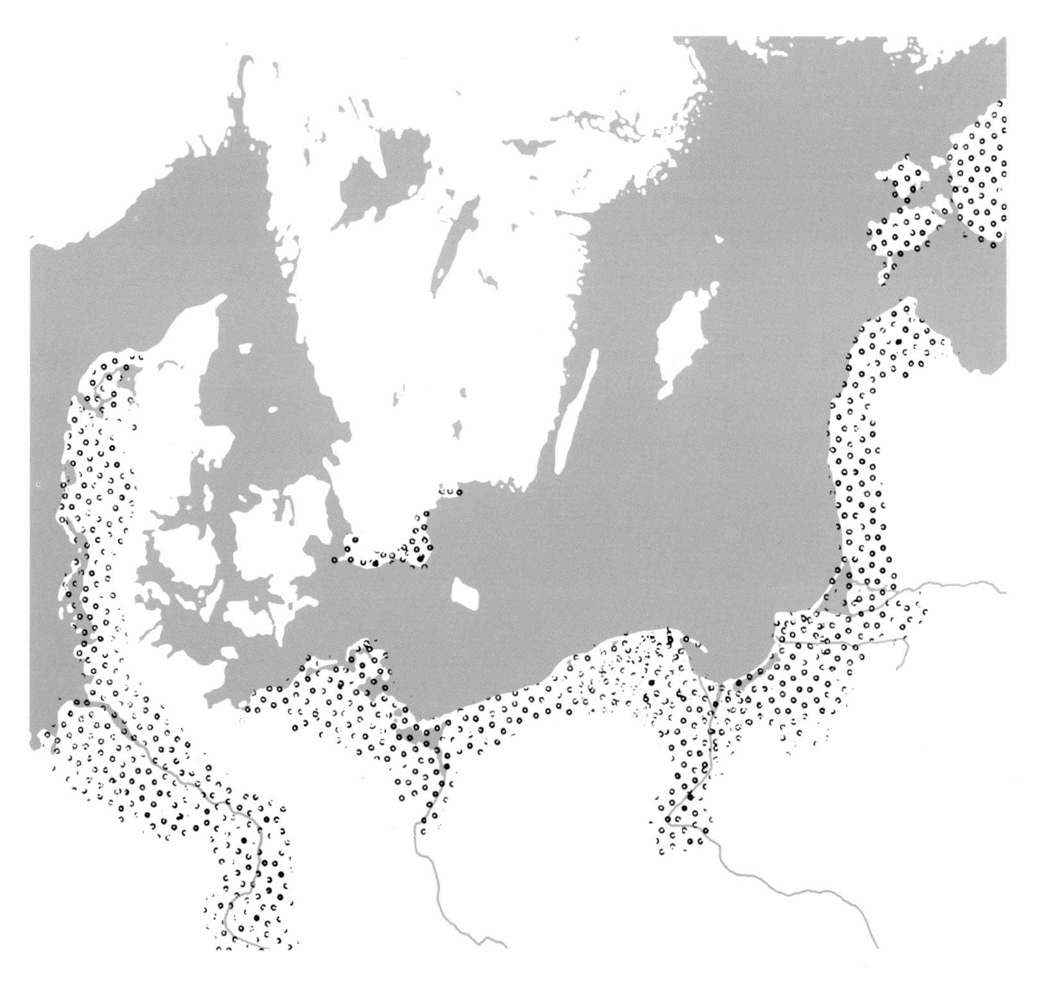

Many stories circulate of fantastic amber finds containing enormous stones. As most fishing tales, they are often exaggerated. A story from Samland tells of a find weighing more than 4,400 pounds (2,000 kg), collected during a single night after a fall storm. That was in 1862. Another story tells of a second tremendous find, weighing 1,910 pounds (868 kg), made in 1914. And in Denmark, finds weighing well into the hundreds of pounds are rumored to have been made, although it is not clear exactly when or where.

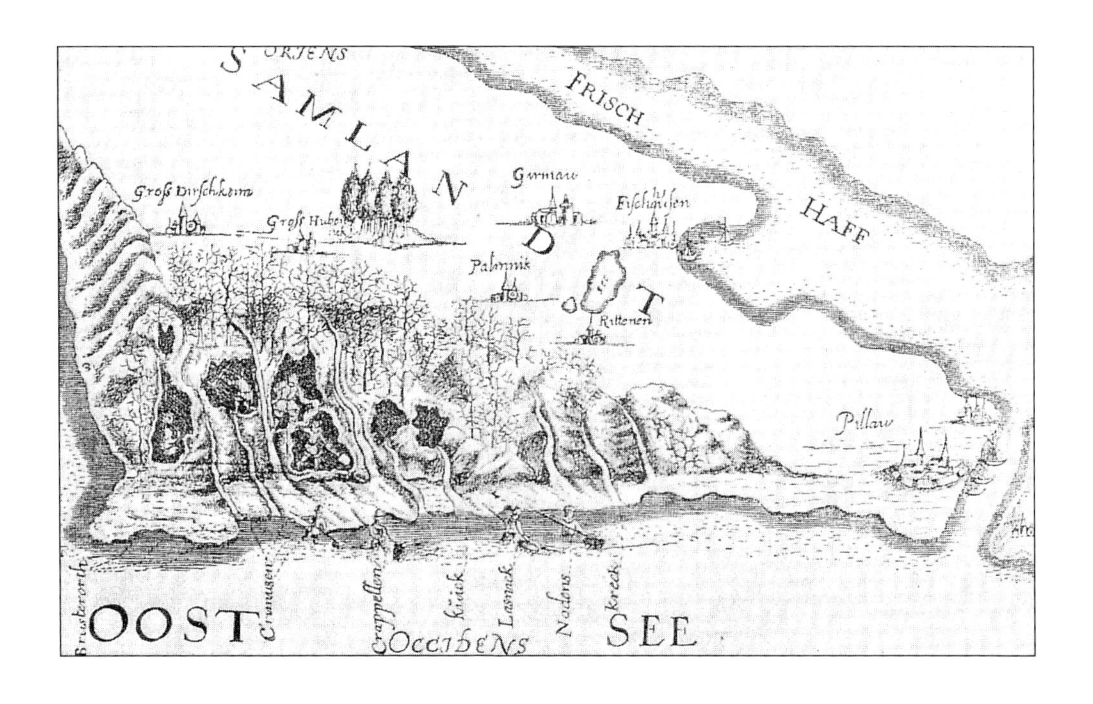

The Alnarps is a prehistoric river containing amber deposits as deep down as 230 feet (70 m).

In addition to the Alnarps's ancient river valley, finds have been made in boggy areas near the coasts of Hanväng and Örnahusen, in the sandy fields of Skanör's Moor and other scattered fields along the coast. As a result of ages of fertilizing the earth with seaweed, amber may be plowed up in many of southern Sweden's old farming fields. The amber, which often clings to seaweed along with other stones, remains in the earth long after the seaweed decomposes. Even Olaus Magnus tells of such finds, located in fields tens of miles from the coast.

The primary sites of amber finds in Skåne, the southernmost region of Sweden, are located in the Kämpinge cove and the Falsterbo headland. Even parts of western Skåne, mainly the stretch between Mîlle and Landskrona, are old amber beaches, as is the southern stretch between Kåseberga and Trelleborg. The east coast has also received a good amount of amber from the sea, but mostly small pieces are found there today.

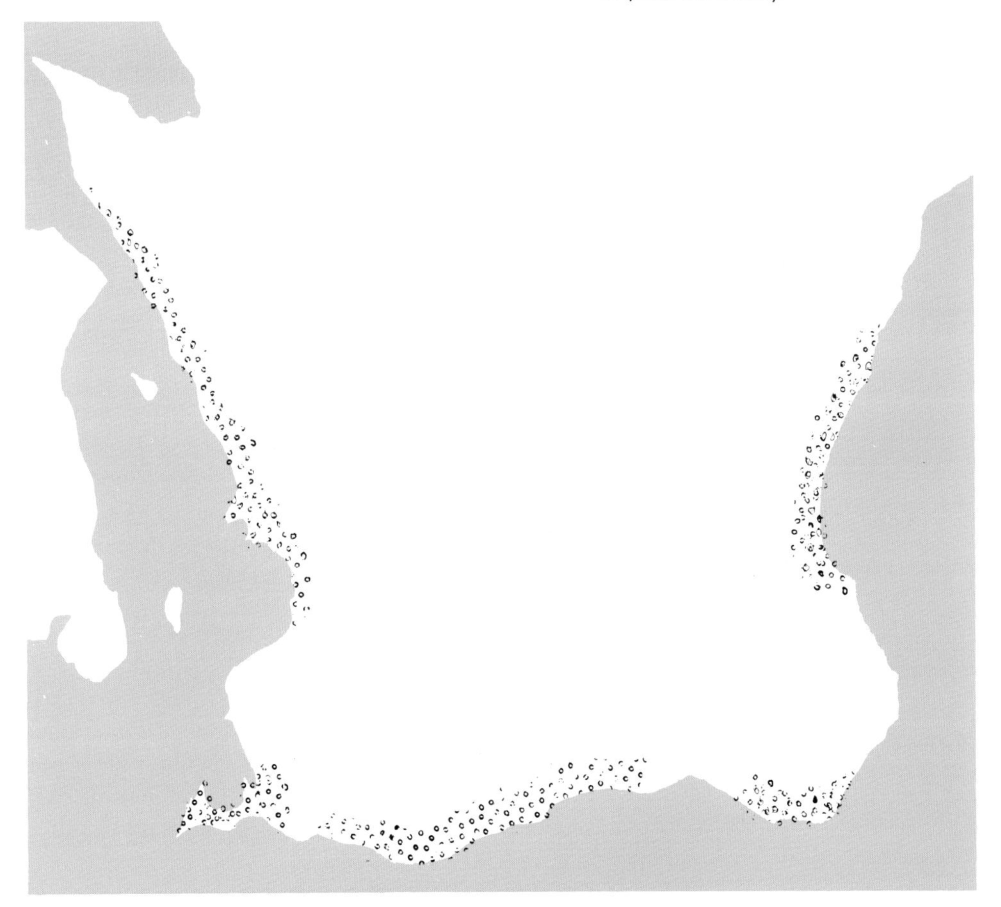

Amber of all kinds

Amber exists, or has existed, in many parts of the world. The total number of amber find sites is around 200. But not all amber is of equal quality. The age and origin of amber vary, as do its chemical properties. The standard means used to differentiate between the varieties of amber has been to analyze its succinic acid content. This compound is most likely the result of microorganisms causing the cellulose in the amber to ferment. Succinite contains 3–8% succinic acid, more than any other type of amber. Some varieties of amber contain no succinic acid at all. These varieties go under the common name of "retenite," and are presumed to be the product of some type of deciduous tree. They are more brittle than the ones containing succinic acid.

This information is useful for, among other things, tracing the origins of the amber artifacts that have been found at various points in the world, and thereby tracing the ancient trade routes.

Analyzing the acid content in amber found in the grave sites of antiquity, researchers have found that it contains succinite, and therefore, they conclude, it comes from the Baltic regions. This method has been criticized as uncertain because other amber varieties can contain high percentages of succinite as well. Recently, however, more reliable methods of analysis have come to be used, such as spectroscopy and mass spectrometry. These methods, which have by and large confirmed earlier conclusions about the amber's origins, are also important to paleontologists, who obviously wish to know where the animals and plants they find in amber come from.

Of the many varieties of amber that exist besides succinite, we will be considering only the ones of primary importance. The most familiar finds are located in Italy (Sicily), Romania, Slovakia, Ukraine, the Dominican Republic, Lebanon, Burma, China, Japan, Russia (including Siberia), Mexico, Canada, and the United States. Amber

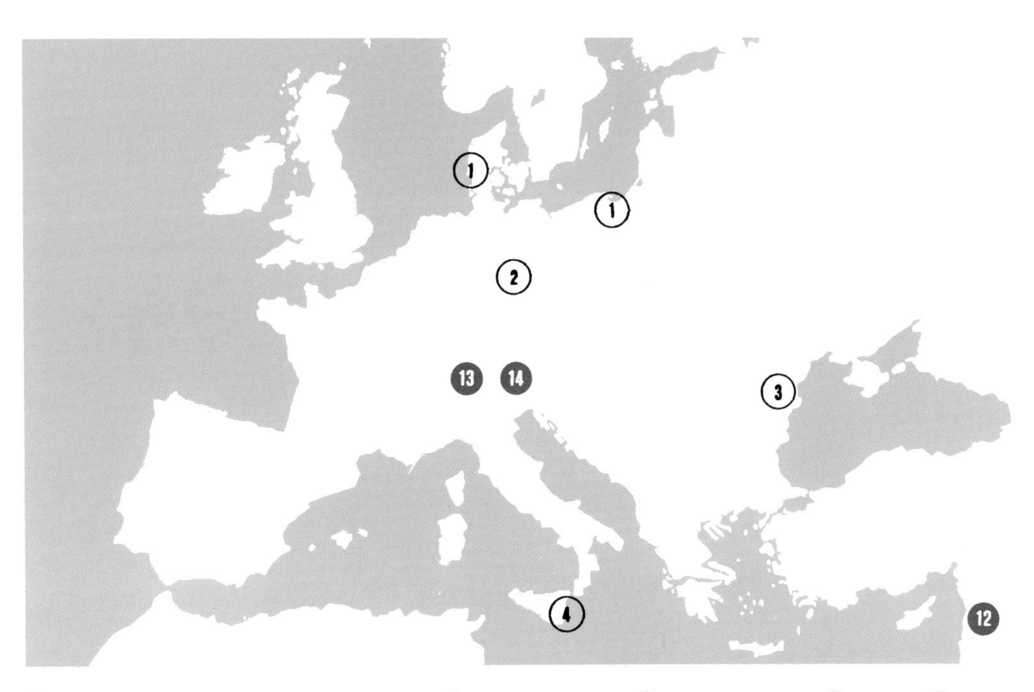

Amber from the Tertiary period has been found in: (1) *Denmark and Russia* (2) *Germany (Bitterfeld)* (3) *Romania* (4) *Italy (Sicily)* (5) *China* (6) *Burma* (7) *Japan* (8) *Borneo* (9) *Dominican Republic* (10) *Mexico*

Amber from the Cretaceous period has been found in: (11) *Canada* (12) *Lebanon* (13) *France* (14) *Austria* (15) *Russia (Siberia)* (16) (17) *United Stated (Alaska and New Jersey)*

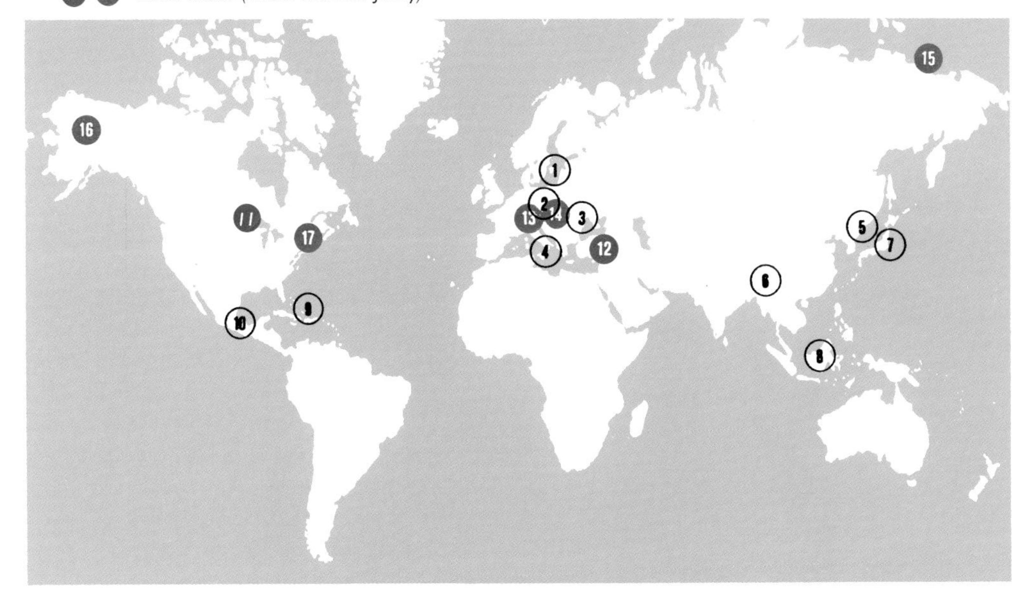

from these areas differs from Baltic amber in many respects. Apart from a few minor finds on the Danish island of Bornholm, which are thought to be 170 million years old, amber from Lebanon—along with some from France, Austria, Switzerland, Siberia, the United States, and Canada—is the oldest in the world. Formed during the time of the dinosaurs, over 125 million years ago (see the geological timetable on pages 124–125), it is rare, however; specimens of scientific interest that are found these days are small in size. As with all amber from the Cretaceous period, it is rather brittle and often contains cracks.

According to a recent publication by George Poinar, amber has now been found amber in 300-million-year-old pit coal. If the amber is from the same period, which would be expected, we are getting closer to the oldest point where amber could exist. Plants did not establish themselves on land until around 400 million years ago; and the history of resin-producing plants does not stretch back more than 360–380 million years.

Poinar has also disclosed findings of fossils of algae, and protozoan specimens like amoebas, in 225-million-year-old Bavarian amber. They can't be seen with the naked eye, but they are there, and they are the oldest amber inclusions found thus far in history.

Simple amber charms crafted in the Dominican Republic

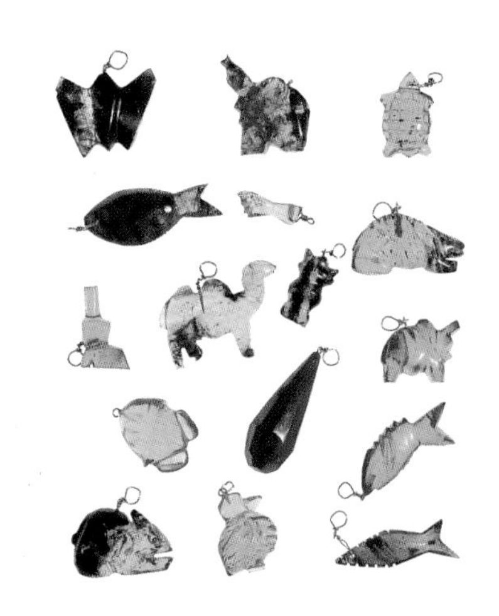

The Dominican amber, embedded in sandstone and dating back 20–30 million years, is often clear and fluoresces in beautiful yellow, red, green, and at times blue color tones. It is also more likely to contain inclusions, the rate being close to 1% of all Dominican stones. Entrapped insects are usually bigger than those in other parts of the world as well, the largest example being a modern-sized dragonfly with a wingspan of 2.3 inches (6 cm). Other examples of Dominican inclusions include termites, cicadas, praying mantises, and scorpions, as well as certain vertebrates such as small frogs, geckos, and iguanas. In one case, regarding a gecko inclusion, scientists used X-ray examinations to conclude that the specimen's backbone had been broken in several places, perhaps the result of a violent struggle

against the sticky hold of the resin. The variety of species in Dominican amber is great; up to 30 different animal species may be found in a single stone. For some reason, perhaps because of the structure of the original resin, the animals' internal organs are better preserved than in, for example, Baltic amber. Plant materials such as leaves, flowers, and root fragments occur more frequently in Dominican amber as well.

The Dominican stone is more brittle than its Baltic counterpart and therefore more difficult to craft. Extraction is hard work; it must be done with a pick and a shovel because dynamite would destroy the fragile stones. The reserves are seldom accessible, and it wasn't until the 1930s that serious extraction began taking place. But it has been successful, and the Dominican Republic has produced some of the most beautiful stones in the world. According to a newspaper article, sizable stones have also been found, one weighing as

In the Dominican Republic amber is found in mountain craters at an elevation of 3,300 feet (1,000 m). The stones are embedded in seabed sediments that have been overturned several times and ended up at the tops of mountains. The three primary points of amber extraction are located in the La Cumbre region, the Palo Alto region near Santiago, and the Bayaguana region. The oldest stones come from La Cumbre and are around 40 million years old. The youngest stones come from Bayaguana; they are less than 20 million years old and in some cases have to be categorized as copals.

much as 13 pounds (5.9 kg). Columbus arrived on this island (then called Hispaniola), and sources tell of his trade in, among other things, amber.

Amber finds in Mexico are located primarily in the east of Chiapas, especially in the Lacandón jungle. Organized mining and gathering does not occur to any great extent. Most of the amber is collected by individuals who search the riverbeds, fields, and mountain slopes or who extract amber from mine shafts in the mountains. These shafts may be several hundred yards long. It is often after landslides and avalanches, when the lignite is overturned, that the "*ambareros*" make their best finds.

Antique amber may still be found in the markets in Burma, but the mines are no longer open.

The Native Americans in the area have long used amber for earrings, nose rings, and necklaces. Crafting the stones is a domestic chore, performed mostly by the women. That the art of sculpting is truly ancient is proved by amber jewelry finds in Mesoamerican pyramid grave sites. It is also known that the Aztecs, like the German Order (see page 59), held a monopoly on amber trade. The Mayas used amber as incense.

As far as we know, Mexican amber is not derived from conifers but, like the Dominican amber (see page 36), comes from *Hymenaea* trees within the legume family.

Amber from Burma (*burmite*) has been extracted primarily in the northern parts of the country and has mainly found its way to China, where it was greatly valued. It is often clear and transparent, but it does not compare to Baltic amber in range and brilliance. One of the largest pieces ever found is thought to be composed of burmite. It weighs 33 pounds (15 kg), and may be seen in the Natural History Museum in London.

Burmese bottle in the shape of the Buddha

Siberian amber is from the Cretaceous era and is rich with insect inclusions. It is found in the Taimyr area but also as far away as Kamchatka, where the locals used it as incense.

Arctic amber may be found, not only in Siberia, but in Alaska and Greenland as well. It is also from the Cretaceous period and is thus at least 65 million years old. In Alaska most amber, or "*omalik*" as the Eskimos call it, has been found on the Alaskan Peninsula, in the Yukon Delta, and on the Fox Islands, which are part of the Aleutians. It is found primarily on beaches and near the mouths of streams and rivers. As one might expect, amber specimens from these northern areas are of great scientific interest, especially if they contain inclusions.

The Eskimos used amber as good luck charms for the hunt—perhaps they still do—and the women wore it as jewelry on their clothes and, it is said, as amulets hanging between their legs.

Canada has around fifty amber find sites. Most of the amber is from the Cretaceous period and contains many inclusions. Two major find sites are located by Cedar Lake and by Medicine Hat and Grassy Lake in Alberta Province. Amber may also be found on Ellesmere Island and Axel Heiberg Island in the northern Arctic Ocean. These finds are the northernmost in the world.

The largest amber finds in the United States have been made in New Jersey. Recent excavations by David Grimaldi have yielded a substantial amount of amber from the Cretaceous period containing many insect inclusions, some of insects now extinct. Flower inclusions and at least one well-preserved mushroom exist as well. The finds are a gold mine, or we should say amber mine, for scientists.

Other amber find sites in the United States are located in Tennessee, Mississippi, Arizona, Nebraska, and North Carolina, as well as along the Atlantic coast in Maryland and on Long Island, Staten Island, and Cape Cod.

One of Linnaeus's apprentices, the professor and traveler Carl Peter Thunberg, reported that he had seen amber in Japan during his trip there in the 1770s: "I was given amber by my friends here. It is called '*nambu*' here, and ranges in color from light to dark yellow. Some stones are striped, and they are said to exist throughout the country."

Thunberg was right. Amber can be found in

(Continued on page 133)

When we think of Baltic amber, we assume that most amber comes from the sea. But the majority of the amber and amber products sold on the market are extracted from the land. As early as the 1930s, up to 90% of the production of Baltic amber was mined.

Palmniks (amber fishers) on the coast of Samland

Fishing, seaweed, and rolled-up pants

Amber fishing from a boat. Drawing from 1868.

You can fish for amber. The phrase may seem curious, but if we can fish for shrimp and other animals then why not for amber? In Samland, amber fishing was an important business. Locals would use either bottom nets, or, if on a small boat or standing in the water, boat hooks and landing nets. Amber could often be found in seaweed that had been uprooted by harsh storms. The amber would come in with the surf and be pulled out with the undertow, if it didn't get stuck or thrown up on the beach. The Samland residents would therefore watch the sea closely, and when the call "It's come!" was heard all the villagers would run down to the shore to salvage as much amber as possible before the drifting islands of uprooted seaweed would float away to competing villages downshore. The men would work in the surf with their landing nets and collection baskets fastened to their waists. The women and children scanned the shore, searching the sand and tufts of seaweed. Fires were raised on the beach, to dry wet clothes and warm freezing limbs, for it was often during the cold parts of the year that the amber-carrying storms set in.

This scene may seem to be made for an artist's eye and brushstroke. But it was hardly a romantic life. Nature could be vengeful, and drowning accidents were not uncommon. Wet clothes became heavy and pulled people down, and it wasn't always enough to be tied to a pole on the shore.

Sometimes people searched the shores before the sun came up. If they walked barefoot along the beach, they could feel the difference between amber and other stones.

In the eighteenth century, some more or less unsuccessful attempts were made at collecting amber with professionally equipped divers. Dredging was successful for a short while, but ended as economic returns waned.

On the North Sea coast, where the tides are pronounced, certain people would ride their horses on the exposed sea bottom at low tide to areas they knew harbored amber. These "amber horsemen" had to be careful to return before the high tide came flooding in. Others accepted the

In the late Middle Ages, Baltic Sea fishing communities were not allowed to dispose of their amber themselves. Under the German Order, during the thirteenth century and a few centuries thereafter, by royal law, all found amber belonged to the Crown (for more on this, see "The Middle Ages and a bit beyond"). The fishermen were given gold and salt, later only salt, for the amber they turned in. Almost certainly, a great deal of smuggling occurred, despite the severe punishment for such offenses—most often hanging. As reminders of the harsh law, gallows were placed at various points, and the king's patrolling coast marshals, who watched the shores day and night, were highly feared. Those who did not live in the area were forbidden to visit the beach. This law remained in force until 1885.
Today tales of ghost riders live on. It is the coast marshals, who have not found peace in death, that have returned to haunt the beaches. They are the Amber Coast's answer to the Flying Dutchman, and on stormy nights they can be heard patrolling the beaches and crying out into the darkness. Or perhaps it is the sea goddess Juraté, who weeps to see the fragments of her amber castle wash ashore (see page 67).

Amber fishermen and a gallows on the Amber Coast. Copperplate print by J. Wagner, 1744.

challenge on foot, and quite a challenge it was. Because of the anxiety and danger involved, they were called "nerve runners" ("*Hitzläufer*").

In our time, the "real" amber collectors are those who, through ongoing observation, know the patterns of amber within the patterns of nature. These patterns can vary strongly due to wind direction, currents, and the topography of shores.

I haven't heard of dogs or seals being trained to search for amber, on land and at sea, but it would not be impossible. It might be nice for weary amber collectors to just let the hounds out in the morning and have them go off to find and collect the amber.

Amber searcher with water glass by Falsterbo lighthouse, one of the best amber sites in Sweden

In the Baltic countries, it is told, some amateur collectors once mistook pieces of phorphorus for amber. The phosphorus came from ammunition that had been dumped into the sea after the war and that had rusted apart. One of the collectors, a boy who put a piece of the phosphorus in his trouser pocket, suffered severe burns on his thigh. A certain degree of caution is apparently needed while beachcombing the seashore for anything that shimmers.

Large amber pieces from the coast of Skåne in southern Sweden

Extraction

Today's largest finds involve land amber. It wasn't until the eighteenth century that this extraction began on a large scale, and it started in Samland. At that time open pit mining was utilized in combination with tunnels dug from the shoreline into the blue earth. These days the amber-rich earth is extracted with steam shovels and pressurized water. The amber pieces are then sorted mechanically according to size and manually according to class. Output varies considerably. Yields of over a million pounds (500,000 kg) per year have occurred, but during both world wars extraction nearly ceased. The mining tunnels were even filled with water during World War II, but the Russians drained them shortly after the war ended. A large portion of the extracted amber is made up of small pieces and otherwise inferior stones. The small pieces are used for producing reconstituted amber, or "amberoid" as it is also called (see page 109). The inferior stones are used in the production of amber derivatives such as amber oil and amber varnish.

In Poland, where amber deposits lay a few yards beneath the earth's surface, the locals would commonly dig small holes, which were sometimes covered with planks, to extract the popular stones. In some parts, the ground became so riddled with holes that the forests suffered serious environmental damage.

Water from a nearby lake is pumped into amber-bearing strata 15–35 feet (5–10 m) underground. The light amber rises to the surface, together with pieces of wood and seashells.

The material that has been pumped up with the water is collected in a net and emptied in a tub of salt water. The amber floats to the surface, while the peat and seashells sink to the bottom.

In moist meadows it is possible to "walk up" amber, either on foot or with livestock. The pressure brings the water and the amber to the surface. An innovation to this technique, discovered by accident during groundwork construction, involves pumping water into soil where amber is thought to lay hidden. When the water and mud are driven to the surface, the light amber stones follow and can be gathered. Here as well, the damage to forests can be too great for these operations to be cost-effective.

New amber deposits are at times discovered on land and at sea during road and waterway construction. This was the case during the dredging of the Kurisches Haff canal in East Prussia, and during the fortification of Copenhagen in the 1660s. Finds are still made today while laying pipes, digging wells, or plowing coastal meadows in amber-endowed areas.

While pumping sand in Kögebukten, Sweden (which once supplied sand for the concrete foundations of the Saab factory in Malmö), amber was accidentally extracted and could be sorted out. This sand-pumping method is often used to reach sunken treasures in shipwrecks or on the sea bottom. If we knew how to pinpoint large amber deposits on the sea bottom or just offshore, this method could become useful for future extraction.

In Bitterfeld, in former East Germany, vast amounts of amber, around 20–22 million years old, have been found under layers of lignite 130 feet (40 m) underground. The stones were crafted into jewelry pieces at the state factory Ostseeschmuck in the coastal city of Ribnitz-Damgarten. The extraction ceased in the early 1990s, and the entire mining area has been converted into a recreation area.

SINGULAR SHIMMER

A piece of amber I take in my hand
from the seaweed on the Baltic strand.

It burns a longing in the skin of my hand—
tell me, stone, what message you bring to this land!

A vision from your singular shimmering light
spellbinds the sea waves' continuous flight.

Is your secret
sealed eternity?

—Wera Engberg
Translated by Jonas Leijonhufvud

Far and near

"Amber Sea" and "Amber Coast"—these romantic names evoke visions that inspire amateurs to take long walks, in summer and fall, along the beaches of Sweden's Österlen and south Skåne. In sand and among pebbles, but especially in seaweed, piles of mussel shells, and other things washed up by the sea, amber may be found. Because seagulls often live near amber, following their flight may be a good idea for the beginning amber collector. Most often you will find only small pieces, but large clumps may be spotted from time to time. It all comes down to having a good eye.

Amber bewitches. Once you have become a collector, it is hard to stop. Some sort of magic must dwell within the stone, for it is seldom that the many long and wet wanderings along the beach pay off at any reasonable rate. A gold digger's spirit, the chance to experience nature, and lots of good, healthy exercise all help out and act as added bonuses. But still, as the Danes say: "First the man takes the amber, then the amber takes the man."

The Swedish writer Jean Bolinder, best known as an author of detective novels, in his novel *Amber Man*, tells about an older gentleman who spends the last years of his life in Sandhammaren searching for, and crafting, pieces of amber. The man called the stone "hardened sunshine from a lost continent." For him, amber was a spellbinding passion. This happens to more than just characters from novels. Many people who have lived near Sweden's amber coasts have become tempted into searching for and sculpting amber. The artifacts range from simple jewels to, for the more experienced artists, small figures and sculptures.

Storytellers have integrated amber into the narrative structures that make stories shimmer and shine. Hans Christian Andersen's little mermaid, who may be seen in statue form on a stone in Langelinie in Copenhagen, lived in a castle with "walls of coral and high pointy windows of the clearest amber." Even the mermaid in Oscar

Wilde's story *The Fishermen and His Soul* lived in a sea palace, "entirely in amber, with a ceiling made from emeralds."

That Bärnsten, Behrnsten, or Bernstein are common last names may be observed by opening a telephone book in Scandinavia, Germany, Great Britain, or the United States. The American composer and conductor Leonard Bernstein, who wrote many symphonies and the music to *West Side Story*, also comes to mind. He clearly possessed some of amber's magic and brilliance; both his music and his personality reflected this. But we can't do much better than a Bernstein symphony when it comes to amber-inspired music. That is, if we don't believe we hear the melody of the sea when we hold a piece of amber to our ear.

THE FISHERMAN AND THE SEA GODDESS

An ancient Lithuanian tale tells of the sea goddess Jaraté, who once lived in an amber castle at the bottom of the Baltic Sea. She was the fairest of all the sea goddesses, and her castle was more beautiful than any other castle, even the ones made from crystal or gold and silver. The sea was her kingdom, and she owned all that was there: the stones, the plants, and the animals.

One day the young fisherman Kastytis sailed out to sea and cast his nets. Since this was in Juratés domain, she became cross and sent her mermaids out to give him a stern warning. But he paid no heed to their warnings, nor to their charms, and he continued to fish. Juraté became even more angered, but she could not help but be impressed with the fisherman's courage. She decided to chase him off herself, but when she saw him, something happened. She fell in love. Instead of turning him away, she took him with her to live in her amber castle.

This infuriated Perkunas, the god of thunder, since he had destined Juraté to marry the water god Patrimpas. In a fit of rage, Perkunas sent a bolt of lightning that demolished the amber castle and killed Kastytis.

To this very day, Juraté still sorrows and weeps, fettered where she stands by the ruins of her palace as punishment for violating the will of the thunder god and loving a mortal. When the storms roar and the waves are high, she is tossed back and forth in her fetters, and through the howling winds and the breaking waves, her weeping cry can be heard. The broken remains of her palace are washed up on the beach, and her tears turn to drop-shaped amber jewels in the sand.

As recently as the 1930s, amber was sold in Swedish pharmacies.

*Recipe for Prince's Drops (*Liquor succinatis ammonici pyroleosus*):*

Acid succinicum depuratum	*100*
Pyroleum animale	*3*
Supercarbonas annonucus	*As much as needed*
Aqua destillata	*As much as needed*

The amber acid is dissolved by heating it in water eight times its weight. Ammonium carbon in combination with purified bone oil is later added to the mix in an amount totaling half the weight of the acid. The liquid is then stabilized with an appropriate amount of ammonium carbon. After a day the liquid is filtered and mixed with one thousand times its weight in water.
The mixture is clear at first but soon becomes light yellow. Later on, it takes on a darker hue, gains a burnt smell, and produces a neutral or slightly acidic reaction.
Should be stored in sealed containers and protected from light.

—The Swedish Pharmacopoeia, 1901

A protection against evil and disease

Both medicine and magic have attributed healing and protective powers to amber. The Romans were convinced of its effectiveness against stomachaches and sore throats, and the women wore amber necklaces to protect themselves from goiters. It was also considered a useful antidote for snakebites, and if pressed to the chest of a wife, it worked as a truth syrup regarding unfaithfulness. Amber was even considered effective against impotence. It eased the pain of childbirth and healed open wounds, although the last may appear contradictory because amber has long been known to delay the coagulation of blood and has therefore been used to make the receptacles and tubes used for blood transfusions.

Amber has also been used against toothaches, even in our own century. In Samland, a home remedy was to hold a piece of heated amber against the aching tooth. It was essential that the stone be thoroughly heated, to a point approaching the patient's maximum tolerance. Young children with toothaches were calmed with amber pacifiers.

Old drawings show us that when epidemics like the plague and cholera swept Europe, doctors used amber incense to ward off disease. In some places, smoking an amber pipe is still considered a protection against infections. Martin Luther, who suffered from kidney stones, was given an amber cross by a fellow believer.

Many native African tribes, and other peoples as well, believed that amber could protect against "the evil eye." Ole Worm, physician to King Christian IV of Denmark (reigned 1588–1648), claims to have witnessed a piece of amber breaking into pieces when an evil man picked it up. Sorcerers used magical amber mirrors to look into the future—perhaps as a way of tinting it with gold. Spindles for spinning wheels were often made from amber, to ward off the witches who would otherwise tangle the thread. If a woman's name was Ann or Anna, it was essential for her to wear amber jewelry so that she would be guaranteed protection from disease and evil spirits. An amber amulet gave both male and female hunters good luck. Sailors were protected from the wrath of the sea by amber amulets or by carrying pieces of amber in their pocket.

The belief in amber's healing and protective powers was more common on the continent, especially in Germany, Denmark, and the Baltic countries, where amber was more accessible, than in Sweden.

Perhaps all this about the healing powers of amber isn't mere superstition after all. Some modern doctors say amber, because of its porosity and millions of tiny bubbles, absorbs the poisons excreted by the skin (although there is no scientific proof that it does). When you hear about this, the saying that white amber, which contains more tiny air bubbles than the other varieties, is most effective against disease makes some sense.

In Skåne, a region of southern Sweden, it used to be said that if amber was rubbed against a pant leg it could be used to draw the dust from your eyes, and there is some sense to this. The people of Skåne also noticed that amber could attract perspiration. It was perhaps for this reason that Frédéric Chopin used to stroke an amber necklace before playing at his piano recitals. Leonard Bernstein, whose surname is German for amber, is said to have used a baton with an amber handle. Perhaps he felt a certain affinity to the material.

It was also believed that amber, if dissolved in vodka, could prevent hardening of the arteries, as well as migraines, for which the temples were dabbed with a cloth dipped in the solution. Against heart disease a rather exclusive recipe was used. It contained white amber, red coral, crab's eyes, cloudberries, pearls, and black crab's claws. A doctor and "scientist"—from the seventeenth century—made it known that poison, if stored in a container of amber for over three hours, would lose its harmful effects. Reliance on such protective qualities—as against snakebites—is obviously misplaced, and experiments should not be attempted.

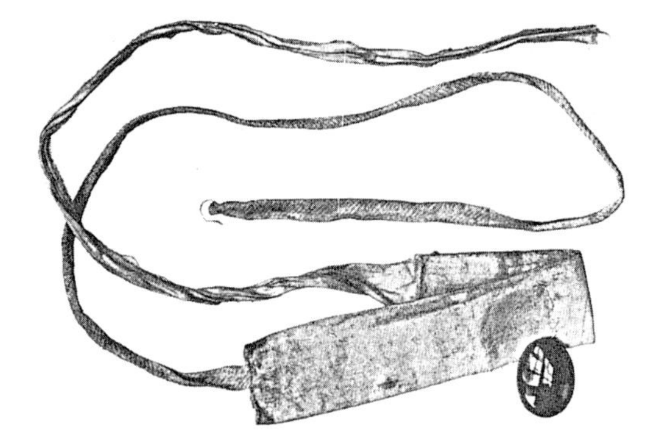

The Romans consumed amber after crushing it and mixing it with honey and rose oil. Much later, amber derivatives, such as amber (succinic) acid and amber oil, were used to make both home remedies and modern medicines, most notably against epilepsy. A medical work from 1672 reads: "For the shivers, take 9 drops of amber oil mixed in vodka." Amber has also been used against gout in a potion called "Prince's Drops" *(see page 68).* *Synthetic amber acid is still used in certain medicines, but the acid exists naturally in certain varieties of mushrooms, berries, and rhubarb.*
Balsamus succini, *or amber balsam, was recommended by medical doctor Johanne Chesenecophero Neritio in his handbook* Regimen Iter Agentium, *printed in 1613 in Stockholm: "Here is a bit of instruction on what to do if you intend to travel, by water or by land, and wish to avoid difficult and fierce diseases. Amber balsam is good for bruises, hits, the falling sickness, and other discomforts of the sea. Place a piece in your ear, and another, the size of a hemp seed, in your nostrils."*
While on the subject of amber derivatives, we might note that the great old violins from Cremona (Stradivarius, Amati, and Guarneri) in many ways depended on the use of amber varnish for their tone and luster. This type of varnish, which has also been used on furniture and finer boat decks, is produced from low-quality amber, amber fragments of no commercial use, and scraps from amber crafting.

Myth and ancient history

Excavations of ancient residential sites and graveyards reveal that amber was valued as far back as the early Stone Age, when it was used for amulets and other types of jewelry. Naturally, examples dating from the late Stone Age are more frequent, and if we examine the evidence from antiquity we find historical documents about amber's uses and functions. It is safe to say that amber has been part of human culture for almost as long as we have existed, thus making it the oldest precious stone still in use.

The American author Jean M. Auel, who researched the Stone Age from many perspectives in order to write her novel about the Cro-Magnon girl Ayla, writes in one of her books about how cherished these magic stones really were. One of the tribes had a summer dwelling place where amber, highly valued in trade with other tribes, could be found.

Examples of historical finds exist in Britain (at Stonehenge), the Baltic countries, Germany, Poland, Italy, and Egypt. The famous amateur archaeologist Heinrich Schliemann (the man who rediscovered Troy) found amber beads and amber jewelry mounted in gold in one of the Mycenaean kings' tombs. Analysis has shown this amber to be Baltic.

Amber finds from the Stone Age in northern Europe tend to consist of beads, amulet stones with holes in the center, small ceremonial axes, and crudely carved figures. Many of these amber figures, despite being crudely carved, display the artist's sophisticated comprehension of form and proportion. The figures, often representing prey such as bear, elk, and boar, also reveal an aesthetic understanding. At times, symbols have been carved into the figures as well. What purpose these carved figures served is not known. Perhaps they were meant to bring luck to the hunt or were offerings to the animal spirits, or perhaps they were simply objects of art.

5,000-year-old amber amulet in the form of a roughly crafted male face with a beard, found in a grave in northern Åsarps, near Västergötland, Sweden

Excavating and dredging in the Baltic countries, in places such as Kurisches Haff, has yielded amber artifacts primarily from the late Stone Age. Besides cylindrical and button-shaped beads with holes drilled into them, crafted figures have been found as well, some in human form. They often have an owl-like appearance, or wear an owl mask, and it is believed that they represent a deity of some sort. It is known that an owl goddess was worshipped by different peoples during the Neolithic era, or perhaps "feared" is a better word, for she was the goddess or messenger of death. (The owl was also the hieroglyphic symbol of death in ancient Egypt.) Other figures represent birds, snakes, elks, bears, and porcupines. Many of the figures have holes drilled in them so they could be worn as amulets or in necklaces.

A number of round amulets, often decorated with rows of circular indentations that curve along the periphery and form a cross in the center, have been found and linked to a somewhat later Baltic culture, also from the late Stone Age. The amulets' shape and our assumption that these people believed amber came from the sun suggests a culture of sun worship. Phallic-shaped pendants have also been found—a type of fertility fetish perhaps.

Animal amulet from the late Stone Age. A find from the excavations at Flackarp in Skåne, Sweden

Amber beads were offered to the gods, worn as jewelry, or used for trade. The treasure that was found in Denmark between Randers and Viborg on Jutland most likely falls into the last category. It contained 4,100 amber beads that weighed close to 20 pounds (9 kg). The 13,000 beads found in Salling, Denmark, were most likely used for trade as well. Several large finds have been made in bogs where the beads have been stored in clay pots; Vendsyssel in the northern part of Jutand has provided the highest yield. During the excavation of the chambered barrow in Barsebäck, a Swedish site from the late Stone Age, around 400 amber objects of various types were found. Several amber objects from the Stone Age have been found in Sweden's southern region of Skåne, among others an animal figure from Flackarp, outside of Lund.

As late as the 1980s an interesting archaeological find was made in Niedzwiediowka, Poland. It turned out to be an ancient center for amber carving and sculpting that dated back to 2000 B.C. The find includes amber jewelry, unfinished carvings, and amber scraps, as well as tools made from flint and other stones. Archaeologists estimate that no less than 900 workshops once operated there.

These club- and ax-shaped amulets, from the late Stone Age, are thought to have been worn to bring luck to the hunt.

Distaff found in Kämpinge village in Skåne, Sweden. Amber distaffs are unusual. Perhaps this one was placed in a grave, to be used in the next life.

Amber die from the Viking Age or the early Middle Ages, found during an excavation in Lund, Skåne, Sweden

In addition to those in Skåne, Iron Age finds have been made in the Swedish regions of Gotland, Västergötland, and Bohuslän. They have contained pearls, amulets, distaffs, and gaming pieces. Even in the Swedish regions of Blekinge, Öland, and eastern Småland, finds have been made.

Finds from northern Europe's Bronze Age are rarer, but many gold and bronze artifacts date from this era. Different theories exist as to why this is. Perhaps the most feasible theory is that the tremendous popularity of amber in the Mediterranean countries caused such a demand that most amber was exported, traded for bronze and gold.

In this context a quote by Olaus Magnus, concerning the amber trade of the sixteenth century, is worth mentioning: "In the land where [amber] is generated" it offers ... little enticement, because it is so common, or because its value is low; the foreigner, however, still appraises it highly. ... It is always this way with desire, if it is satisfied too often it causes man to become weary of it; and in turn, objects that are rare and used sparingly, give rise to desire and reverence."

Another reason for the rarity of amber artifacts from northern Europe's Bronze Age is that cremation became standard during burials, and people's jewelry and amulets were destroyed as a result.

Thus also, neither the Iron Age nor the Viking Age has left many amber artifacts in Sweden, although significant finds of raw as well as sculpted amber were made during the excavation of Birka. In the èhus area of southern Sweden, an Iron Age workshop that had made amber beads and other artifacts was found. Nearby, amber amulets in the shapes of Thor's hammers, whetstones, and other things were also found.

Many of these finds are exhibited in the Historical Museum in Lund. A grave find from the Iron Age also exists in Simris, in the same part of Sweden. It is a spool-shaped piece of amber that was placed in the dead man's mouth. The placement, according to Swedish scholar Berta Stjernqvist, indicates that it was considered to be a type of coin and was intended as payment to Karon, the ferryman of the afterlife. The belief in this mythological figure also existed in Germanic cultures.

The Phoenicians, antiquity's foremost merchants and explorers, traded with amber, or "copper saffron" as they called it. They were quiet about where they found the stones but eager to tell tales

The Baltic amber routes mainly were over bodies of water such as rivers and seas. The three main routes were

1. The old amber route, via the Elbe, Saale, Danube, Inn, and Adige (Po) Rivers to the Adriatic Sea;
2. The later amber route, via the Rhine and Rhône Rivers to Marseilles;
3. The most important amber route, via the Vistula, Danube, and other rivers to Aquileia in the northernmost part of Italy.
Beyond this, important trade routes for amber were created from the Baltic areas to the Black Sea. From the same area, amber was also transported on waterways to Finland and Russia.

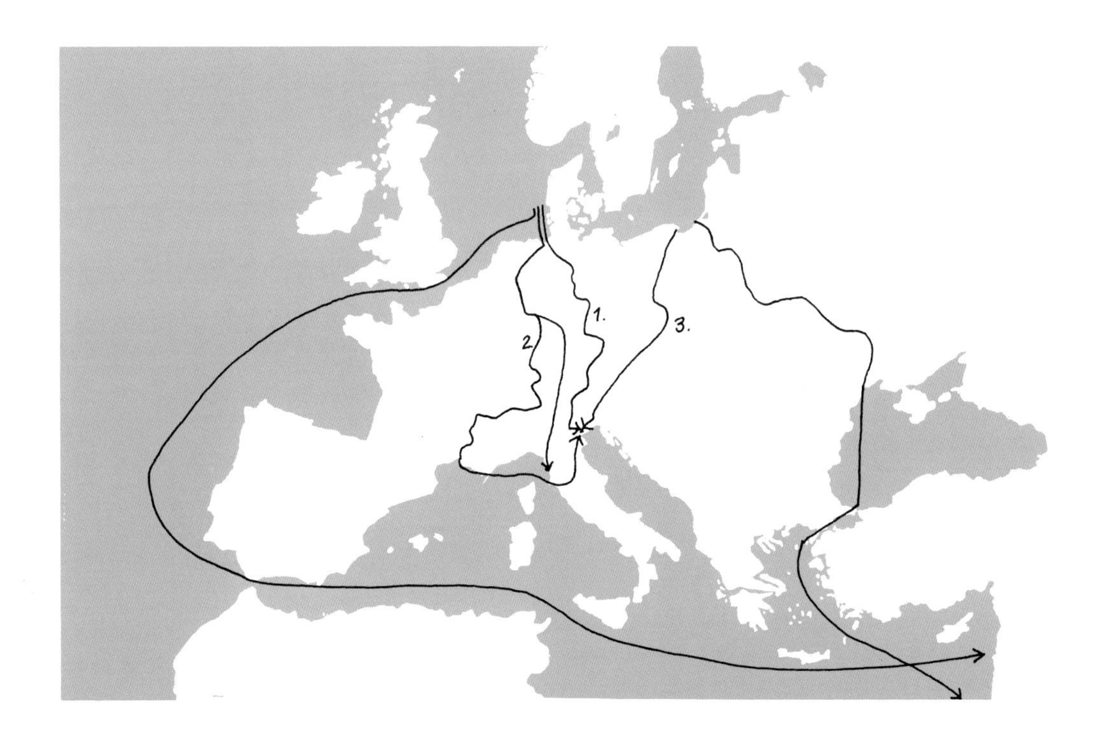

Already in the Stone Age, amber buttons, with drilled holes, could be made.

Detail of a horse statuette from the Bronze Age, with eyes made from amber. Helsinki, Finland.

Whetstone, Thor's hammer, and cat figure, found during an excavation at Birka, Sweden

of sea monsters and other abominations in the sea in order to discourage competition. With time, trade routes were established between northern and southern Europe. It was not without reason that these were known as the "amber routes," even though they served to transport other goods, such as furs and slaves (prisoners of war), that were traded from the North to the South.

Prehistoric trading beads made from amber, stone, and glass

Greeks and Romans

Documentary evidence shows that the Greeks, and to an even larger extent the Romans, valued amber immensely. The stones were used primarily for making jewelry, amulets, and medicine. The *Odyssey* refers to amber jewelry in several places. Penelope, for example, received an amber necklace from a suitor.

In the Roman Empire the city of Aquileia, at the southern end of the Adriatic Sea, was the main receiver of Baltic amber, and even the center of amber sculpting. Many finds that may be seen in the city's museum today confirm this. During Nero's reign, amber was at the peak of its popularity. It was then that the "Amber Knight" set off for the "Amber Sea" in the north to acquire amber in large quantities. The expedition, which went through Donau and Weichsel, was a success and the amber was used extravagantly, among other

A small amber ornament was worth as much as a healthy slave in the Roman Empire.

things, to decorate the gladiator arena, the gladiators' outfits, and the stretchers used to carry out the slain competitors. Nero apparently even used a piece of amber as a monocle—an early example of a sunglass. There is evidence that emperors at times threw amber, instead of gold coins, to the people. A Roman author grieved over the high price of amber—a small, well-sculpted piece was worth as much as a healthy slave.

One of the earliest explorers, the Roman Pytheas, who was originally from Massalias (modern Marseilles), discovered as early as 300 B.C. the "road to the land of tin and amber," that is, the way to Britain and most likely the North Sea island of Helgoland, currently a part of Germany. Pytheas's own version of these adventures has been lost, but they are retold by Pliny the Elder, who describes Helgoland in the following way: "In spring the waves wash up amber on the beaches of this island. The natives burn it as firewood, and sell it to their Teutonic neighbors." The Swedish author Alf Henrikson has written a novel based on a passage of Pytheas's exploration of Thule (believed by modern scholars to be somewhere in the Shetlands, Iceland, or along the coast of Norway) and the Amber Sea.

The Middle Ages and a bit beyond

In the Middle Ages the use of amber for making rosaries became so dominant that amber craftsmen became known to the Germans as "*Paternoster macher*" (makers of Lord's Prayer books). Customers were not only Catholics, but also Muslims, Buddhists, Tibetans, and others. A find of several amber rosaries and uncrafted amber at the excavation of Helgeandsholmen, an islet in Stockholm, suggests that amber rosaries were produced in Sweden during this period.

Rosaries from various religions. Islamic (left), Catholic (middle), and Buddhist (right).

Rosaries are used by, among others, Buddhists, Muslims, and Roman Catholics. The Catholic rosaries are made up of 55 beads, some of which are larger than the others. The largest beads are called the "paternoster beads" (the Lord's Prayer is read to them), and the small beads are called "Maria beads" (Ave Maria is read to them). Smaller rosaries, with 33 beads, also exist, as do larger ones, with 165 beads.
The Islamic rosaries have 99 beads—the hundredth is in paradise.
The Buddhist rosaries have 108 beads, which corresponds to the number of Brahmin present at the Buddha's birth.

On the continent, the German Order gained a leading role in amber craftsmanship during the Middle Ages and a bit beyond. The order was made up of retired crusaders, who after their return from the Holy Land had established themselves as colonialists, missionaries, and to some degree robber barons in eastern Europe, primarily Prussia, Poland, and the Baltic countries. The order was organized strictly by class, with the knights and priests on top. These were recruited from the German nobility and wore white cloaks with black crosses. The leader was chosen by a general assembly, much like a parliament, that included the knights. Their headquarters were eventually located in the castle at Marienburg, today's Malbork, several miles southeast of Gdansk. It was declared that all the amber in the area belonged to the German Order. This shows

what an important trade commodity and income source amber was during this period. No other people were allowed to own any uncrafted amber. Only the head of the order could allow exceptions.

Eventually, these rules were applied less rigidly, and the order employed its own "*Bornstein-sniczer*," or amber lathers, many of them from Königsberg (present-day Kaliningrad). The power of the order decreased with time. After a war lost to Poland in 1466, its former status became but a memory, although the organization was not dismantled until 1525.

In the new Prussian state—initially under Polish leadership—amber craftsmen were allowed to establish themselves. New craftsmen's workshops sprouted up here and there, in Danzig and Königsberg among other places. The production, which had up until the sixteenth century consisted mainly of rosaries, became more artistically oriented, especially with declining demand for rosaries since the Reformation.

The monopoly eventually evolved into private competition, with the exception of white amber by reason of its medicinal properties. Elector Frederick William of Brandenburg, known as the "Great Elector," repurchased the monopoly in 1642, and all amber was once again owned by the state. New laws were adopted. This time it became illegal not only to possess uncrafted amber, but also to walk along the beach without permission. Fishermen and locals were forced to swear an amber oath every three years, promising to turn in smugglers even if they were close relatives. A special amber court was created to administer the oaths and sentence the violators. The punishment was most often hanging. Despite all this, the state monopoly suffered losses and was dismantled in 1811 during the Napoleonic Wars. The obligation to turn in found amber to the government, however, remained in effect until the end of World War II, although the punishments had by then become fairly small and the gallows were no longer waiting.

All amber that was found by the sea was to be turned over to the German Order immediately, which had collection posts at various points and many storage vaults in castle basements. With respect to the law against owning amber, and in an attempt to prevent violations, amber was not allowed to be crafted within the community's borders. It was transported in barrels to Brügge and Lübeck, where guilds for amber craftsmanship were established. (Brügge gained one in 1302, and the one in Lübeck was founded in 1360.) From a prosecutor's standpoint, this was a brilliant move. Persons found with so much as a single piece of raw amber in their pocket could be considered a criminal, while those who wore an amber rosary crafted by guild members were not only good Catholics, but also honest members of society.

The Virgin Mary standing on a crescent moon. Crafted around 1550 in the Polish cloister Oliwa.

In Denmark, a royal decree on amber was issued as well, stating that all amber along the beaches belonged to the king. Marshals guarded and collected amber along the coasts. The stones would sometimes be auctioned out, and the marshals would then receive part of the profits.

Goblet without a base, thought to be made by Sigismund, king of Poland (1587–1632) and of Sweden (1592–1599)

The Reformation

When rosary production decreased as a result of the Reformation, new uses for amber were developed. The real blossoming of amber art took place between the sixteenth and the eighteenth centuries. Many artifacts were made by combining amber with other materials, such as ivory. Centers of production were located in Königsberg,

Sigismund was a powerful supporter of amber craftsmanship, and he was a craftsman himself—although not one of the masters. Some of the artifacts thought to have belonged to him are still in existence; one of them is a goblet currently kept in the Royal Castle of Wawel, in Krakow. Also kept in the same castle is a fine necklace with a medallion that once belonged to Sigismund's consort, Anna, queen of Poland and Sweden.

Engaging in craftsmanship was a popular pastime in the courts; many examples of amber and ivory artifacts remain. Members of the nobility imprisoned for political reasons were often allowed to cultivate such hobbies. For example, the daughter of Danish King Christian IV, Leonore Christine, upon the death of her husband, who had been imprisoned on suspicion of treason, was locked up in Blå Torn (the Blue Tower), part of the old Castle of Copenhagen where state prisoners were kept. There she crafted many beautiful amber artifacts, while Magnus Stenbock, also imprisoned in Denmark, preferred to work with ivory.

Women riding camel in translucent amber on a podium of opaque amber. Poland, seventeenth century.

Crucifix in amber with Christ figure in ivory. From the second half of the seventeenth century.

Danzig, Lübeck, Stolp, Brügge, and to some extent Copenhagen. A number of talented artisans, who developed an understanding for the unique properties of amber, established themselves in these regions. They often had ties with a certain court, or were commissioned by a court to create their artifacts. (Some of the more famous of these craftsmen are listed on page 127.) Much of their work still exists, and although it does not always carry a signature, sources and comparisons have sometimes been used to identify the artifact as the work of a specific craftsman. Other times, it hasn't been possible to attribute a piece to a certain artist, at least not with certainty. As with the work of many great artists it is suspected that certain artifacts, or parts of artifacts, were created by apprentices or students in the master's workshop.

To begin with, craftsmen were restricted by the size and shape of the amber pieces. But later, by layering or integrating amber with other materials, usually wood, large objects such as cabinets and cases could be made out of natural amber. To create smaller artifacts or parts of larger ones, putty was sometimes used to connect fitted amber pieces with each other without layering it onto other materials. In this way, the luster of translucent amber could be appreciated when the light shone through it.

Even though many of the amber artifacts were created in the form of household articles, they were not meant to be used. Examples include bowls, goblets, pots, and plates. Other objects made with great craftsmanship are cases, cabinets, canisters, inkwells, hourglasses, game boards, chess pieces, crucifixes, home altars, candleholders, chandeliers, cases for clocks, frames for mirrors, and of course necklaces and jewelry. Additional amber objects from this period include handles for knives and forks. A beautiful example is a preserved pair of silverware with the carved head of a man (perhaps representing the owner) at the bottom of the knife's handle, and the carved head of a female (his wife?) at the bottom of the fork's handle. The handles are posi-

Goblet made in Poland during the first half of the seventeenth century

Jewelry reconstruction from a residential site. The find was made at the Bo estate in the parish of Bredsätra on the Swedish island of Öland.

tioned so that the carved heads face each other when placed on either side of a plate.

Another slightly odd application was to make glasses, monocles, and magnifying glasses with lenses from clear amber. The castle of Skokloster in Sweden houses a pocket mirror mounted in an amber frame; it was meant to be hung from a band in the lining of a dress. Examples of musical instruments occur as well. Frederick the Great, who enjoyed playing the flute, was given such an instrument made entirely of amber. Amber's light weight would be ideal for making a crown, fit for any king. Such a crown was made for Johan Sobieski III, who was the king of Poland from 1674 to 1696. Unfortunately, it was stolen from the royal treasury and its fate is unknown.

Well known, on the other hand, and highly venerated by the Poles, is the Black Madonna from the Polish cloister Jasna Gora, an altar picture constructed to hold different adornments (such as a crown), which are fastened to it on holy days and certain other special occasions. One of these adornments is made of amber and red coral. Among the thank-you gifts presented to the Madonna for listening to prayers, many are also made of amber.

Even though a great number of the amber artifacts have been destroyed, or disappeared during World War II, and even though many of those remaining have been severely damaged by time and poor handling, there are a significant number of such amber treasures left. Quite a few of these have found their way to Sweden.

So many artifacts were made for the Great Elector in Brandenburg, where amber was regal, the court there came to be called the "Amber Court." Several of these artifacts were used as diplomatic gifts. A good number of them may be seen in Sweden's royal castles, primarily in the Royal Palace in Stockholm and the Castle of Gripsholm, and also in the National Museum of Stockholm, the Nordic Museum, and the Royal Academy of Science in Uppsala.

In China, amber has been appreciated for a long time as well. Many beautiful artifacts and

After the amber embargo ended in the Baltic countries, the local population could once again reap the "sun gold" the sea and land had to offer. Amber is referred to in many ancient Baltic folk songs, and it is still is used to make decorations and buttons for Baltic folk costumes. Amber was often included in the dowries of the women in the coastal regions, or it was given as gifts to the new in-laws. This is illustrated in the following folk song:

I brought to my brother
a wife made of amber.
All of her dowry
shimmered of amber.
She gave father
a shirt made of amber.
She thanked mother
with a shawl made of amber.
On brother's arm
she hung a towel made of amber.
On sister's head
she placed a crown made of amber.

jewels of various kinds were made, especially during the sixteenth and seventeenth centuries, when China experienced a blossoming in amber craftsmanship as well. Most of the artifacts are made from imported burmite. Traditional jade craftsmanship is reflected in the style of Chinese amber art. Most of the pieces are figures, ornamented bottles, bowls, or vases, all in both opaque and clear amber. Beijing is, and has always been, the center for amber trade and production in China. Collections of Chinese amber art may be seen in the West in museums such as the Victoria and Albert Museum in London, the American Museum of Natural History in New York, and the Museum of Fine Arts in Boston.

Canister lid in translucent yellow amber. Danzig, second half of the seventeenth century.

The Amber Room

One of the foremost diplomatic gifts, a chair made entirely from amber, was given to the Russian Czar, with whom it was important to stay on good terms. But a more extravagant gift was to be created, the largest amber work ever to be planned and completed. Called the "Amber Room," it was a large chamber for a royal castle with walls and furnishings almost entirely covered with amber. The vision for this room is said to have been had by Frederick William I of Prussia—although the seed for the idea may have been planted by someone else—and King Frederick IV of Denmark was also involved. In any case the work was undertaken in Prussia by a craftsman from the Danish royal court. The artist's name was Gotfred Wolfram, although he

The world's largest amber artifact. It disappeared without a trace after World War II.

went by the name Gotfred Lacquei. After seven years of working on the project, he got into an argument with the foreman and returned home. The work was then taken up by two craftsmen from Danzig named Turau and Schact. The room was intended for the Charlottenburg castle, at that time located outside of Berlin, and was nearly completed by 1711. It was later added to and moved, first to the State Palace of Berlin, and later to the castle in Potsdam.

When, in 1713, Peter the Great of Russia visited Frederick William I, Frederick I's son, he was so impressed with the room that Frederick William thought it best to give it to him as a gift—an extravagant but perhaps well-calculated diplomatic move. In return, Frederick William, who created the Prussian army and greatly admired tall soldiers, was eventually granted a company of fifty-five soldiers, all over 6'6" tall.

Although the Amber Room's wall panels were broken up into twelve pieces and transported by boat and sled to Russia, the room wasn't reassembled until sometime between 1740 and 1760—sources vary in their accounts. The Amber Room's new home was in the summer castle Tsarskoe Selo located in what is now the city of Pushkin, outside St. Petersburg. It was Peter's daughter, the empress Elizabeth Petrovna—the one fond of men and court life—who oversaw the work. Peter himself had lost interest.

Because the hall in this castle was larger and higher than the one in Charlottenburg, the walls had to be complemented in some way. For this, the Italian architect Bartolomeo Rastrelli was commissioned. He solved the spatial problems by integrating large mirrors and gilded wood carvings, among other things, into the structure. The Prussian regent contributed an amber palate with the symbol of the Czar, a two-headed eagle, in bas-relief. (The original panels were adorned with the Prussian seal, a crowned eagle in profile.)

The walls were now 33 feet (10 m) wide and 15 feet (4.5 m) high. The amber, placed in mosaic patterns in varying shades of yellow and brown, and with engravings of such motifs as sea- and landscapes, was not attached directly to the walls of the room but to large wood panels. The room was illuminated by candelabras and chandeliers with faceted amber stones hanging from them. The other decorative objects in the room were also made, entirely or partly, from amber. Visitors were duly impressed by the grandeur and charm of the room, which was accentuated by the softly tinted light reflecting off the amber. Entering it was, as one visitor expressed it, like opening a jewelry box.

When the Germans attacked what was then called Leningrad in September of 1941, the Russians understood that the Amber Room was in danger. The Germans could easily justify taking it, proclaiming that the work would once again return to its home country, and it would end up decorating the walls of Hitler's palaces. But time was not on the Russians' side. Dismantling the walls would take too long, so they decided on an emergency solution and covered the walls with wallpaper to hide the amber. But fooling the Germans proved to be a difficult task. They knew where the room was located and with customary meticulousness they dismantled the walls and sent them, together with the decorative artifacts, to Königsberg in East Prussia. The entire piece was displayed in the citadel there, to be admired by the Nazi leadership.

In 1944, when the allies began bombing East Prussia, the room was dismantled again and packed into boxes. What happened after that is still a mystery. The person responsible for the treasure after it had arrived in Königsberg was the director of the art museum there, Alfred Rohde, who has also authored some well-informed works on amber. He disappeared after the fall of Germany in 1945 but returned during the Russian occupation. When questioned, Rohde denied all knowledge of where the Amber Room's dismantled walls and artifacts were. Both he and his wife died of dysentery in December of 1945, without providing any hints or clues for an eventual search. The death certificate was signed by a doctor who could never be located.

The Amber Room may be resting at the bottom of the sea inside some torpedoed ship, or it may be stashed in some hidden basement. Perhaps it was destroyed in the war. Perhaps, as some maintain, it is part of the art collection of some wealthy American, or in the possession of some escaped Nazi living in South America. However, many believe that it still exists somewhere within the vicinity of Königsberg (now Kaliningrad). Much time and money has been devoted to searches, and they will most likely continue. Treasure hunters are optimists and don't give up easily. Among the excavating parties were the East German secret police Stasi, who spent hundreds of thousands of marks in their attempts.

The Russians have officially given up hope of ever retrieving the Amber Room. But around the year 1979 an enormous project of recreating the lost treasure was initiated. Its targeted completion date was 1990, but only around a quarter of the work is complete today. Perhaps this shouldn't come as a surprise. The goal was to make an exact replica of the room. This requires detailed studies of the few, unclear photos that have been preserved, whereof only one is in color, and an endless puzzling together of amber pieces of specific size and color. The technique used is the same as the one used for tapestries. That is to say, the photographs are enlarged to the proportions of the wall, and then computers are used to read differences in, for example, shades of color. Limited access to the particular kind of amber needed, and higher amber prices, most likely contributed to the delay as well. The project is now being completed by a private artist's group called "The Amber Room LTD." The group is active in other restorations as well.

A story that keeps surfacing in different versions is that the boxes with the Amber Room in them were loaded on to a ship bound for Germany at the end of the war, and that the Russians torpedoed the ship. When the sunken ship was examined after the war a large hole was found in its side and the treasure was gone. This, of course, opens new paths of speculation.

According to a recent report, the dismantled room may be hidden inside a closed mine in Volpriehausen close to Göttingen in Germany. A telegram that was sent to Berlin from Königsberg at the end of the war, by a certain SS Sturmbahnführer Ringel, may be interpreted to indicate that. During an examination of the mine, archives and an amber collection belonging to the University of Königsberg were found, but there was no trace of the Amber Room. An explosion apparently prevented the search from being completed.

Amber in our age

After centuries of artistic innovation, interest in amber crafting declined and was not rekindled until modern times. During the intervening period, the amber industry suffered some difficulties with declining demand and rising costs. Various attempts were made to stimulate the faltering market, and conventions were organized for amber and amber products. In Germany the Nazis frantically sought new applications for their Prussian amber. Among other things, they began to use amber objects as prizes at sports events instead of the usual trophies made from precious and less valuable metals. Amber prizes were also

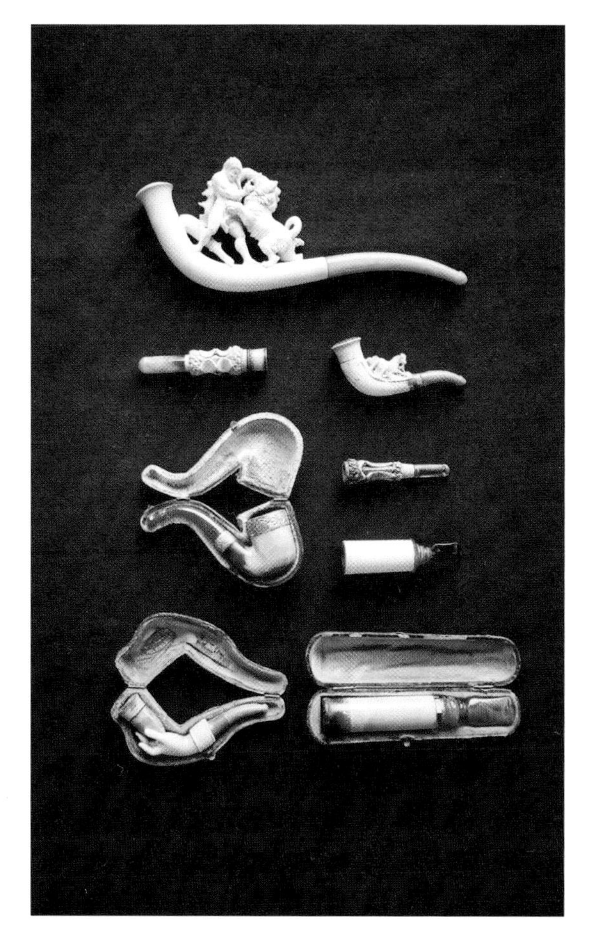

Pipe stems and cigarette holders made from amber. The pipe bowls were made from meerschaum. To drill curved holes in the stems was not possible in those days, so the stems were bent during heating.

Because the amber would melt if it came into contact with burning tobacco, it was not used to make the bowl of the pipe. For that, meerschaum was a better material. Meerschaum bowls and amber stems were long considered the ideal combination for pipes. Cigarette holders made from amber were also popular, with a protective device at the end made from metal or some other material that wouldn't be damaged by the burning tobacco.

Meerschaum pipes

In the old days, meerschaum was thought to be made from hardened sea foam, and the German word Meerschaum *literally means "sea foam." Today, we know that meerschaum is actually a mineral, a magnesium silicate, that is mined mainly in Turkey and Tanzania. The only similarities it has to sea foam are its color and that it is porous enough to float in water. Meerschaum is soft, readily workable, and ideal for making pipes.*

It all began in Budapest. A nobleman who had acquired a piece of meerschaum from Turkey and didn't know what to use it for turned to a shoemaker, who had carved as a hobby, and asked him to try to carve a pipe out of it. The shoemaker crafted two pipes, one for his customer and one for himself. But before he smoked his pipe, he accidentally smeared some shoemaker's wax on its stem. He noticed that the pipe gained a beautiful amber color where the wax had stuck. So he waxed down the whole pipe. Then he noticed that not only did it look better, but the smoke tasted better as well. That was the beacon signal. Meerschaum pipes and cigarette holders quickly became fashionable in Europe, and production was centered in Vienna. Eventually, pipe bowls made from waxed meerschaum evolved into fine artifacts and became symbols of prestige. They were carved into heads with beards, ancient gods and goddesses, scenes of St. George confronting the dragon, and so on. The finest pipes had stems made of amber, but other materials were used as well. Despite the fact that meerschaum pipes grow more beautiful the more they are used, the most expensive ones were probably taken out of their cases only on special occasions—that is, if they weren't permanently kept on display behind the glass of a cabinet. The Tobacco Museum in Stockholm has a fine collection of meerschaum pipes and of cigarette holders made from meerschaum and amber.

given out at the 1936 Olympic games in Berlin.

Most of the commercially produced amber products were made by the company Stantien & Becker, which long dominated the market. For a while, various forms of amber tobacco accessories such as pipe stems and cigarette holders were popular. Amber was said to have a cooling effect on smoke, making it easier to enjoy.

Today jewelry completely dominates the amber industry. But amber canisters, napkin rings, paper knives, seal handles, and clock faces are also common, and an increasing number of decorative ornaments are being sold. Motifs such as flowers, mushrooms, and animals are often sculpted from a single piece of amber. Larger objects such as decorative windows, lamps, and boat models are usually made by combining amber of various colors and types.

The most frequent types of amber jewelry are necklaces, armbands, medallions, pendants, brooches, and earrings. Because of its softness, amber is not well suited for rings, but it may, of course, be mounted in rings made from other materials such as gold and silver. The material's lightness is an advantage, especially regarding earrings. The necklaces may be made from unsculpted amber, machine-tumbled, rounded pieces, hand-cut and polished pieces, or—most beautiful and precious of all—faceted beads. Examples of men's jewelry include cufflinks and tie clips made from amber and usually mounted in silver.

Unique stones are worth their weight in gold; most pieces are priced just above the rate for silver, while the price of treated amber varies greatly.

A truly fine necklace, made from evenly colored natural amber, is difficult to produce because the color and density of amber varies greatly from stone to stone. A large selection is needed to yield comparable stones. Through using color additives and annealing the amber, nature may be circumvented, and almost any density and color combination may be produced. A trained eye, however, will easily detect when this has been done.

Amber model of the ship Santa Maria. The Amber Museum in Ribnitz-Damgarten.

Especially eye-catching are the models of old sailing ships with masts, rudders, sails, flags, armaments (such as cannons), lanterns, and figureheads, all done in amber. The models are made by hand. They demand that the craftsman have a keen eye and a steady hand. Access to amber in a variety of colors and transparencies and a good amount of time and patience are also required. The body is usually made of flax wood, on which more or less flat pieces of amber are glued. For the parts where rounder pieces are required, such as the sails that need to look like they are swelling in the wind, the amber is worked out and cut into appropriate shapes. One beautiful such ship was made by Gunnar Bodin in Skanör, Sweden. The largest amber ship ever made is said to be the model of the frigate Wappen von Hamburg, made by the German Alfred Schlegge. It is 59 inches (151 cm) long and 51 inches (130 cm) high. Eighty-eight pounds (40 kg) of amber, and three years of work, went into building it.

Stones with inclusions are especially popular for pendants and brooches. Because the supply of these stones is quite limited, their price is much higher. It is also common, through heating the amber in sand and then chilling it, to create patterns in the amber reminiscent of inclusions. In actuality it is the air and water bubbles that have been altered so that they look like fish scales, tiny parasols, leaves, and such. These patterns are often referred to as "sun spangles."

In various parts of the world, but primarily in Russia, alternate uses for amber and amber products are being researched. One area of potential application is lasers. Another is electronic components, where amber's electrical properties may come to a broader use. Perhaps even medical science will find new uses for amber in the areas of medicine and artificial limbs.

Modern uses differ from those of the past, and today no one suggests burning amber as incense or fuel, to provide illumination, or to repel mosquitoes. The accounts of the Roman explorer Pytheas's journey to the Amber Sea in the fourth century B.C., as retold to us by Pliny the Elder, reveal that the natives of the North Sea island of Helgoland were in fact burning amber for fuel. This may seem extravagant to us, like setting fire to money. It was not done to impress, however, as was the case in ancient China, where the rulers would set fire to amber during special celebrations.

The amber on today's market comes mainly from Russia, Poland, and the Baltic countries, which also possess the greatest supply. Amber has always been an important commercial product. An example of this was when the Lithuanian amber cooperative, Ambra, became the first to be allowed to privatize after perestroika and Lithuania's subsequent liberation.

Reports have come out of Russia and Poland indicating that organized crime has begun playing a part on the fringes of their amber industries. History seems to repeat itself, as the harsh tactics of the German Order are echoed in today's criminal organizations.

Silver weddings and gold weddings are familiar terms, but that an amber wedding should be celebrated on the tenth day of marriage is a more recent and less known concept. But why not? An amber necklace, given on the tenth day as a protector and talisman, can be as good a proof as any of warm devotion, and perhaps it can provide the magic formula that keeps the relationship together.
In any case, an amber necklace is light and easy to wear and always feels warm against the skin. But it could also be a little risky. According to old folklore, an amber necklace would choke anyone who told a lie while wearing it.

Amber craftsmanship

The interest in crafting a piece of amber you have found has always been great, especially because this may be accomplished with a few simple tools. To succeed, you must have a good technique, an artistic eye, and a feel for the material. But when these elements come together, your imagination is the only limit.

In crafting amber, it is important not to violate the stone's natural qualities. Each piece of amber should be judged individually, according to its shape and structure, before its application is determined. The sculpting process should not be taken too far. The stone's iridescence is often concentrated near the surface and leaving some of it uncrafted often creates a nice effect, even on cut and polished stones. Simplicity in design and craftsmanship is also something heavily stressed in the only existing school for artistic amber crafting—located in the Latvian city of Liepaja. A Polish amber artist echoes the philosophy in the following statement:

What can be more beautiful than a piece of captured sunshine that has received only minimal polishing, so that its natural charm is kept intact, and that has been`mounted in silver, a metal that perfectly complements the color and structure of amber?

In its simplest form, the crafting process may be accomplished with a hacksaw for the rough work, a knife and some sandpaper for the more delicate tasks, a drill for hole making, and a polishing cloth together with precipitated chalk, toothpaste, car wax, or cigar ash for the final polish. The stone should be hand-held while crafted, not put in a vice; otherwise the brittle amber might easily break into pieces, although professional amber craftsmen use machines to cut, lathe, polish, and tumble the stones. In the old days, converted foot-driven spinning wheels were used, but now production involves electric machines. A dentist's drill with a whet plate attachment has proven to be a perfect tool for both rough and fine work.

These "sun spangle" inclusions are not natural. They were created through "clear boiling," or during heating. The faults, or cracks, are often formed around air bubbles or areas of microscopic contamination. Through knowledge and experience, it is possible to affect the size and the number of the sun spangles.

This technique does produce quite a bit of amber dust, and craftsmen should use some sort of protection to avoid breathing it in.

Hole making is often a tricky process and is best done with the amber piece placed under water. This avoids making cracks in the area where the drill goes through.

Tumbling is done basically in the same way that rock collectors do their delicate work. The stones are placed in a motor-driven tumbler together with beach sand or a special powder. They are tumbled for a few days, and then retumbled with a polishing substance. Rounded beads for necklaces can be made in this way. Faceted beads need to be made manually, a process that increases their price significantly. The color and luster of the finished product is best brought out by rubbing the stone with a cloth dipped in vegetable oil.

If a pattern is engraved on the back of a piece of transparent amber, it appears as if it were inside the stone, a technique often used in glass sculpting. By filling the grooves of the engravings with gold or silver foil the artist may achieve remarkable effects.

In a French variation of this technique often employed during the Reformation, amber was "*eglomisé*," or combined with other materials into single artifacts. This process involves coating the bottom of transparent amber disks with, for example, gold foil, and then engraving figures, letters, or ornaments in it. Afterward, the engraved areas are filled in with an appropriate color. The disks are then fastened to the object that they are meant to decorate, a case, a board game, or something of that nature.

As we have seen, muddy and opaque stones may be made clear by slowly bringing them to a boil in vegetable oil, and then just as slowly cooling them down. Color may be added during this process. Today, amber is first heated in an autoclave together with nitrogen to become clear, then heated in an oven to gain a reddish brown hue and the previously mentioned sun spangles, which are produced when the air bubbles explode and the stone gains "substance."

Industrial-scale craftsmanship occurs mainly in Russia, Poland, Denmark, and Germany. Located in Königsberg (now Kaliningrad), the most important factory employed no less than 1,500 workers in 1914—1,000 on the floor and 500, mostly women, working in their homes. The latter were given a certain quantity of raw amber, to be crafted according to certain specifications. They were expected to return not only the finished product but also the scraps. To prevent theft, the total amount was weighed in and compared to the weight that had gone out.

Under the Communists, Russian amber was sold through the state-owned company Almazyuvelireksport and Polish amber through the state-owned company Cepelia. Today, trade has been liberalized. Polish amber, for example, is often distributed and sold by traveling salesmen, tourists, and fortune hunters. The products are not always of the best quality, and sometimes they aren't even made from amber (more on this in the next chapter). So be careful!

Olov Andersson, Falsterbo's most famous amber craftsman throughout the ages

The Collectors

Like many parts of Denmark and elsewhere around the Baltic Sea, the coastlines of Sweden's southern region of SkÅne have been home to many, more or less regular searchers and collectors of amber. Many of the people who lived by the beach were looking primarily for driftwood, but they kept an eye out for pieces of amber at the same time. Others looked only for amber, and many of these crafted the stones they found into artifacts and jewelry pieces. For most, it was a kind of hobby, but that did not prevent them from selling some of their creations. As a result, the sale of amber jewelry and artifacts was a popular and picturesque feature of the marketplaces in Skanör, Falsterbo, and Höllviksnäs. It still occurs sporadically during the summer.

There has always been competition between the amber collectors. Many have their favorite spots, and they keep them secret. They feel imposed upon when amateurs and others encroach on these. Today, there is no one who resorts to such measures as old "Knössa Jeppson," who used to shoot recklessly with his shotgun at people entering certain areas he saw as part of his personal domain. On the other hand, a recent story tells of a contemporary amber collector who would carry amber fragments in his pockets and spread them around on certain parts of the beach after a storm. This would mislead amateur collectors and allow him to search the fruitful areas with less chance of disturbance.

Most often, only small pieces of amber end up on the beach. The larger stones have to be collected in the water, during low tide, in holes and hollows that are sometimes referred to as "amber nests."

Professional collectors use nets and rakes. Often these tools will be connected to either end of a single staff. Some use boats and water glasses (see page 60); others use diving gear and swim to the places where they think they can find amber.

To separate amber from seaweed and other materials common to the beach, the material to be searched is placed in a container with salt water. The heavy objects then sink to the bottom, while the amber and other light materials float on the surface. The surface material is then gathered and placed in a container of fresh water. This time the amber sinks and can be separated easily. This method, used primarily in Canada and Poland, is more time-consuming than simply looking through the materials and is therefore appropriate only under certain conditions.

It is possible to dig for amber if you know just where to look. According to Olov Andersson, however, in Falsterbo, Skåne, it doesn't matter where you dig. Because much of the region was once below sea level, amber can be found in almost the entire area.

The best finds on the coast of Skåne are in the south.

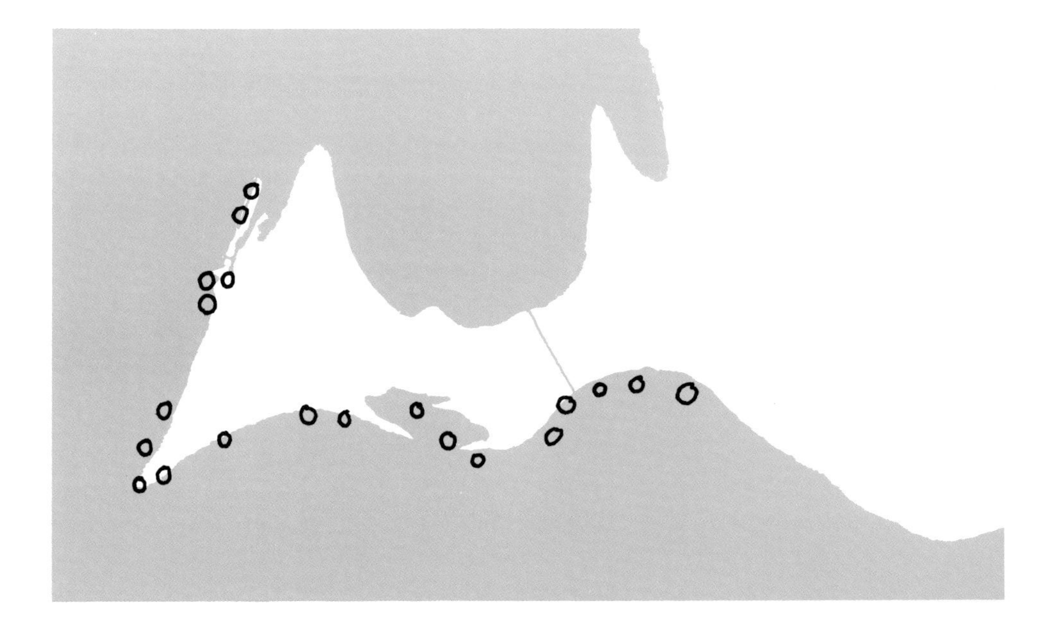

The Craftsmen

The first prominent Swedish amber craftsman to admit to his skill was a grain merchant and alderman from Trelleborg named Johannes Borgman (1816–1884). With a few simple tools, many of which have been saved, he would sit and cut figures out of amber in his free time. He sculpted miniatures of boats, wagons, anchors, cliff rocks, barrels, and tools. Today most of his work is kept in the Malmö Museum, but some of it is also kept in the Falsterbo Museum. Unfortunately, time has not been kind to these artifacts. A well-known piece is a diminutive glass bottle with a chain containing 365 links, one for each day of a year, connected to its cap. What patience and skill such a work must have called for! Other artifacts by Borgman worth mentioning are a scale, a broadax, a cross of the German Order, a miniature flagpole, and many medallions and small pictures. One of the pictures reads: "Have patience. Be pious." It seems to reflect his own virtues quite well.

Another well-known Swedish amber craftsman was Olov Andersson. After working as a baker for thirty years, he returned to his original home, a cottage near the coast between Falsterbo and Skanör. As a twelve-year-old, Andersson was already herding sheep on the heather-clad moors of Skanör. But he still had time to look for amber along the beach. Eventually, he came up with the idea of carving figures out of the soft stones. As with many innovators, Andersson was laughed at, but he kept with it. He changed professions, however, and for a while his job was to wake up the citizens of the town of Falsterbo every morning. This was done by walking through the town's streets dressed in a special cloak and blowing a loud horn.

It wasn't until after Andersson left his bakery job, however, that his amber visions could be realized. Together with his wife he began collecting amber on his coastline property and in the surrounding areas. From his finds he made necklaces and other pieces of jewelry which he sold to both local and foreign tourists. He was eager to tell that his customers included wealthy women from America, Paris, and even Japan (at least in one case). He also had students, most of them tourists as well, who had found pieces of amber on the beach and wanted to learn how to craft them. He taught them not to overwork the amber, but to maintain the stone's natural character and form—good advice that still holds true.

Olof Andersson died in 1967.

Even today there exists a dedicated enthusiast, who collects, crafts, sells, and exhibits amber. He is coauthor Leif Brost, from Kämpinge, and he is the only person in Sweden today who calls himself a professional amber craftsman. Since he was young, Brost has collected amber and accumulated knowledge about the stone in his native Sweden and around the world. He is an artist and an adventurer in the most positive sense. He and his wife Lena have traveled around the world visiting places where amber can be found. His collection includes amber stones, inclusions, jewelry pieces, artifacts, and amber imitations from all over the world. Of his amber artifacts and jewelry pieces many are from ancient times. His exhibition in the House by the Sea Museum in Kämpinge fishing village is well worth a visit. The area alone is impressive.

Much evidence indicates that a Viking trading post was located near Kämpinge —Skåne's answer to Birka or Hedeby. By observing the environment and studying finds from excavation sites, and by cross-comparing this information with ancient writings, old maps, and tales out of oral tradition, Brost has found clear indications of ancient Viking culture in the area. Hopefully, archaeologists will follow up on these finds. If an ancient trading post did exist near Kämpinge, and amber could be found along the coast, then there is good reason to believe that the valuable trading commodity of amber was crafted in the region.

In recent years, Brost has become internationally known in the amber business by holding exhibitions in places such as Berkeley, California.

In Sweden, he has participated in exhibitions at the Natural History Museum (Naturhistoriska Riksmuseet) in Stockholm. It is not without reason that journalists call him the "Amber King."

Even though much of the amber collected along the beaches of Skåne by locals and tourists remains hidden in old drawers and cabinets, most of it has been bought up and exported. According to Swedish scholar Carl Sahlin, it was primarily migrating Jews who acquired the amber in trade for the products they were peddling. Amber was also bought at markets, mainly at the market at Kivik.

Some of the purchased amber was sold to larger distributors in the coast cities, or directly to craftsmen and goldsmiths. Some of it went to pharmacies who sold it as medicine in both solid and liquid form. An example of the latter was a potion called "Prince's Drops" (see page 68), which remained in the Swedish pharmacopoeia until 1925, according to Sahlin.

Amber was also sold to Denmark, and Danish buyers even advertised in the local newspapers of Skåne. But much amber, especially the small pieces and the craftsmen's scraps, was exported to Königsberg in Germany, where it was used in the production of amberoid or melted down to produce amber derivatives.

Today raw amber is bought only by craftsmen. Some of them have regular suppliers, who systematically search the beaches.

The world's largest piece of amber?

In early 1988 a sensation was announced in the amber industry. The world's largest piece of amber was thought to have been found. It weighed 525 pounds (238 kg), a gigantic piece when compared to the previous record holder weighing a mere 33 pounds (15 kg). The newspapers wrote articles on it, and Leif Brost went to have a look at it. The story of this event is told below.

As usual when a trawler's net is hauled in, there was excitement in the air that day in February 1988. Jan Ekstam and his crew on the boat Kapduva were fishing in the Baltic Sea southwest of the Swedish island of Gotland. A disappointed sigh and a few loud oaths may have slipped Ekstam's tongue when the crew reported: "Large stone in the net." This happens now and then, and there is always a risk that the net will become damaged. The only consolation was that the journalist and photographer on board the ship would be able to witness the hardships and risks that accompany fishing at sea.

Slowly, the net was maneuvered aboard, until the stone could be examined above the water. There was something strange about it. It didn't look like a normal rock; it sparkled and shimmered here and there. It appeared to have broken in two—a similar, smaller piece was seen falling back into the water. Luckily, the large piece remained and was carefully brought abroad.

The skipper had seen something like it before, something much smaller. Could it be a piece of amber? In that case, what a boulder! The world's largest piece of amber, up till then, weighed 33 pounds (15 kg), but this must weigh more than twenty times that amount. What could it be worth? A million dollars, or maybe more?

The journalists would put it on the first page. The world's largest piece of amber has been caught in a trawler's net! But first, an examination was conducted. A jeweler examined a few fragments of it. Unfortunately, it wasn't genuine amber, but perhaps it was fossilized resin, maybe a million years old, or maybe only a thousand years old, or perhaps younger still, or perhaps it was hardened amber, or copal varnish from a barrel that had gone overboard a ship at some point in time. The excitement died out, the hope for the million dollars vanished beyond the horizon, and the work on Kapduva returned to normal. But it was exciting while it lasted.

Leif Brost, however, was still interested in the find. He spoke to Jan Ekstam via boat telephone and arranged to purchase it and have it taken to Gräddö, an island that served as Kapduva's home port. The boulder was still in the ship's storage room and three men were required to move it into a car. While it was being moved from the car, a few more pieces came off. The boulder is now in four parts, tied together with sturdy rope, with the largest piece weighing 300 pounds (137 kg).

The boulder is currently exhibited in the amber museum in Kämpinge. Brost has had it examined in the Quaternary geology laboratory at the Lund University, where carbon-14 testing was conducted. The resin—they called it that—according to the test, dates "from anywhere between the seventeenth century and 1950. A certain (very small) possibility that the sample is from the sixteenth century also exists." Which doesn't tell us much. But at least the resin boulder made for a good story, even if it didn't end the way we might have hoped. It should be added that just under a dozen similar resin boulders, weighing between 10 and 110 pounds (5 and 50 kg), have been found. Most were pulled out of the Baltic Sea, but a few were found in the North Sea as well.

Genuine or fake?

There is artificially created amber, there are amber imitations, and there is fake amber. In the nineteenth century, the process of using small pieces of amber to create large ones was invented. The method involved heating the amber under high pressure in containers that cut off all air supply. The compressed amber, which is called "amberoid" or "mosaic amber," is genuine amber, but it isn't natural amber. It contains shifting, often darker, color tones, and it lacks the luster of natural amber. By artificially adding colors, however, the hue may be altered. Amberoid is not valued as highly as the natural kind. It is commonly molded into rods that are later cut up into smaller pieces and used to make beads and other objects.

Telling the difference between amberoid and natural amber may be difficult for the layman. A slightly filmy appearance and a symmetrical form, without any irregularities in color or structure, are warning signs. If air bubbles occur, they are usually oblong, as opposed to the ones in natural amber, which are usually round.

On the other hand, amberoid remains the stronger of the two materials and is often used to make objects that require durability. Examples include handles for canes and umbrellas, doorknobs, door handles, cigarette holders, and pipe stems.

This boulder of hardened resin, brought up in the net of a Baltic Sea trawler, weighs 525 pounds (238 kg). Most likely it was part of the cargo of a wrecked merchant ship. The resin was probably stored in barrels and used as raw material to make pitch or varnish.

Another product that is sold as amber, and in a way is amber, is *copal*, sometimes called "African amber." Copal is also a hardened resin, but of a much younger vintage. It has been found, in some abundance, in Africa, south and east Asia, South America (mainly in Colombia), and New Zealand. It is not as hard as Baltic amber and does not have its luster. It also smells differently, and if held in the hand, it begins to get sticky after a while.

The copal, whose name comes from the Nahuatl (Aztec) word *capalli* meaning "incense," is referred to as a "semifossil" in the industry. It may be millions, thousands, or merely hundreds of years old. The older and harder it gets, the more like amber it becomes. One of the oldest kinds of copal comes from Mizunami in Japan. It is said to be around 330,000 years old. Raw copal, tapped straight from the tree, may also be bought on the market. The trees may be conifers, as is the case with the well-known kauri copal from New Zealand, or deciduous and belonging to the legume family, as is the case with the African Congo copal.

The kauri pine (*Agathis australis*), which is not a pine but a member of the Araucaria family, is a magnificent tree, up to 165 feet (50 m) tall, and a national symbol of New Zealand. Because of its valuable wood, the tree has come close to extinction, but some stands remain and new trees are being planted. The tree releases a resin that hardens into copal, which may, after a few million years, turn into amber.

The natives of New Zealand's islands, the Maori, used the copal for chewing gum and to clean their teeth with. They also discovered that, much like tinder, the half-fossilized copal they found in the earth could be used to start fires. The smoke from the copal also kept the insects away. In addition, burnt copal was useful for making tattoos. When the Europeans arrived in the 1820s, they used the copal for fuel and, dissolved in oil, as varnish, which eventually made copal an important export product.

When the aboveground deposits had been used up, digging for copal began. Many European immigrants tried their luck as copal diggers and collectors. Often large boulders were found. The largest weighed 185 pounds (84 kg), and is displayed at the museum in Dargaville. As the demand increased and the supply decreased, people began tapping the living trees for raw copal instead. They cut holes in the tree trunks and allowed the resin to seep out and harden into lumps. After a few months the lumps were chopped off, and the cuts were exposed again. This treatment was made illegal after many trees died. Today, the industrial demand for copal is quite low. As usual, synthetic substitutes have taken over, but because copal varnish dries faster than its synthetic counterparts, it is superior under certain circumstances.

New Zealand copal found in the earth is between 850 and 45,000 years old. For some reason, natural inclusions of animal and plant parts are rare. On the other hand, faked inclusions are widely marketed.

Genuine amber may also be found in New Zealand, but not in any large amounts. The deposits that have been found are mainly located in coal seams and are at least 20 million years old. It is surprising that the amber finds are so few in relation to the finds of semifossilized copal, especially because large forests would have covered the islands during the Jurassic and Cretaceous periods. The upheavals in the earth's crust during the Carboniferous period and the Oligocene epoch, and the ice movements during the ice ages, may have carried the amber away, or buried it in places where it is difficult to find. Perhaps the amber is now at the bottom of the sea, or in some other country.

In connection with this, the question is raised of exactly what should be categorized as amber. Should copal count as amber and be allowed to go under that title? The Germans have instituted regulations for amber trading to the effect that any products under 20 million years old are not allowed to be called amber. In that case, copal is out of the picture. In Sweden, we don't have such rules, but in professional trading essentially the same standards apply. Whatever dating threshold is used, however, it is important to be careful when buying amber, whether on the open market or from individual vendors.

Attempts to produce amber synthetically have had little success. In the few cases where success has been claimed, the cost of production has been much too high to be commercially viable.

As with most things that are desirable and rare, imitations have been made from cheaper materials. Many types of imitation amber exist, and many more are sure to come along. In some cases the products are openly declared as imitations, while in other cases they are conscious counterfeits.

Synthetic resin, plastic, and even glass have all been used to counterfeit amber. Also available are products made from a combination of plastic and amber—with the plastic usually composing the larger portion. Amber powder has also been mixed into the plastic imitations, in order to mislead the customer who conducts a scent test. With necklaces and armbands, it is sometimes the case that only some of the beads are genuine, while the rest are made from plastic.

Another way of cheating is to fake inclusions. This has been done with natural amber as well as amberoid, not to mention the various synthetic materials. A hole is made in the object and a fly or leaf is placed inside, and the hole is then filled in with the same material the object is made from. Copal may be melted and poured over the "inclu-

Amber imitations made from plastic are sold as "African" amber in antique and mineral conventions all over the world, but they are most commonly sold in the tourist bazaars of Africa itself.

sion." The nonexpert won't be able to tell the difference between a fly from the Tertiary period and a fly of today. So if an offer is made to purchase a "unique" inclusion, it may be a good idea to have it examined first. Particular caution should be taken if the inclusions are of larger animals, or if the insects seem to rest peacefully in the stone, without any signs of struggle.

But it isn't always so easy to tell the fake from the genuine inclusion; even the experts are sometimes fooled. It was recently discovered that a famous inclusion of a common latrine fly (*Fannia scalaris* Hennig), which is part of an exhibit in the Natural History Museum of London and which has been examined and classified by insect and fossil researchers, is a counterfeit from the nineteenth century. At the turn of the century, J. P. Morgan paid $100,000 for an amber collection that he gave to the American Museum of Natural History in New York. The collection's main attraction was an amber piece with a frog inclusion. It wasn't worth the price. The inclusion was soon found to be a skillfully made counterfeit. Swindlers have always been around, and amber counterfeits date back as far as the sixteenth century. In Arthur Conan Doyle's *A Study in Scarlet*, Sherlock Holmes exposes a near-industrial counterfeit operation busy at work inserting artificial flies into fake amber.

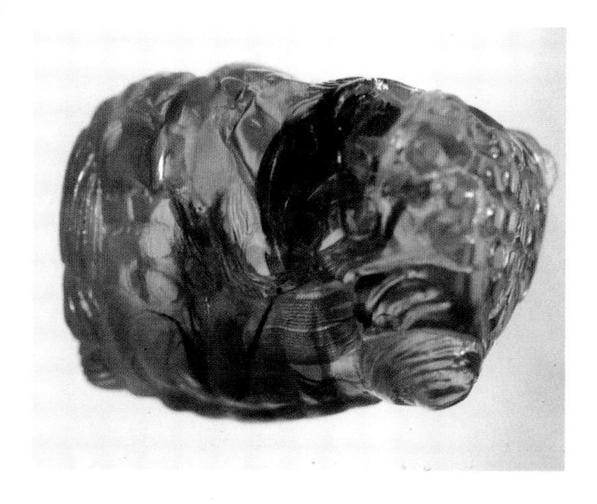

Plastic "fossils" were mass-produced in China after the success of the film Jurassic Park.

In this context, it may be worth mentioning that a synthetic resin now exists that may be placed around an object you wish to preserve. In this way, insect collections may be made that won't be damaged by air, moisture, vermin, or negligent handling. Even Canada balsam, made from the bark of the North American balsam fir, may be used for this purpose, as it is for mounting and sealing microscopic specimens on slides.

To complement and elaborate on the above-stated information, the most common imitations and their properties are outlined below.

Imitations

Celluloid or *nitrocellulose,* the first real plastic, was invented as early as the 1860s, perhaps with the main intent of imitating ivory. As imitation amber, it is called "*ambre antique.*" It smells like camphor when ignited. Today, celluloid is seldom used, much because of its flammability, and other materials have taken its place. But "celluloid amber" still exists in small parts of the markets.

Specific gravity: 1.26
Refraction index: 1.60–1.66

Bakelite (phenol formaldehyde) was invented in 1909 by the American chemist L. H. Baekeland and named after him. The product is sold in powder form to producers of imitation amber, who cast them in certain shapes. In the 1920s, many "amber necklaces" were produced from various colors of bakelite—red was most fashionable. Bakelite is harder than amber and emits a sharp, pungent odor when ignited.

Specific gravity: 1.33
Refraction index: 1.55

Casein, one of the chief proteins in milk, has been used very little to make imitation amber, despite its ability to assume a natural amber color. Its main use has been in the manufacture of buttons.

Specific gravity: 1.33
Refraction index: 1.55

Polystyrene is a thermoplastic appropriate for casting. Lighter than amber, it has become quite popular among the producers of imitation amber. It dissolves easily in gasoline or toluene.

Specific gravity: 1.05
Refraction index: 1.55

Plexiglas (polymethyl methacylate) is a plastic with a high potential for creating amber imitations. But it hasn't been used to any large extent.

Specific gravity: 1.18
Refractive index: 1.50

Bernit, or *Bernat,* is a German synthetic strongly reminiscent of amber. It also contains "stress spangles" that emulate the natural crack formations found in genuine amber. Bernit with insect and plant inclusions may be found, although it is obvious these have been inserted recently. The air bubbles found in most genuine amber in conjunction with insect inclusions, usually around the insect's breathing organs, are missing in Bernit imitations, as are the traces of the insect's attempts to escape.

Specific gravity: 1.23
Refraction index: 1.54

Polybern, another product developed in Germany, is composed of polyester in combination with small pieces of genuine amber. The name reflects this: *poly* stands for "polyester" and *bern* stands for "*Bernstein,*" German for "amber." It is one of the most advanced imitations to date, and it is difficult to see the difference between it and genuine amber. In one version of this product, the plastic contains amber dust, which makes it smell like amber when ignited.

A Polish product similar to Polybern contains less amber and has mainly been used to make necklaces. For a while, it flooded the market.

Acrylic compounds are used in the United States in "Slocum imitations." Produced by Slocum Laboratories in Royal Oak, Michigan, these may be purchased in cubes made for sculpting. They come with or without insect inclusions and smell like burnt fruit when ignited.

Specific gravity: 1.17
Refraction index: 1.50–1.55
Hardness: 3.00

Glass can be recognized by its higher weight and hardness. It also feels cold to the touch, unlike genuine amber, which always feels warm. Colored glass is usually used to produce the imitations.

Animal horn, although gray in its natural state, may be colored to look like amber. Used mostly to produce beads, it has a higher specific gravity than regular amber and smells like burnt hair when ignited.

Testing authenticity

There are many ways of testing the authenticity of amber. Genuine amber floats, or hovers, in saturated salt water, unlike glass or artificial resin imitations. Plastics, on the other hand, are light and will even float in water with a low salt content. When amber is rubbed with a wool or silk fabric, it becomes electrically charged and will attract small bits of paper, something glass and copal cannot do.

The characteristic fragrance that amber emits when ignited may also be used to test its authenticity. To avoid damaging the stone, a heated needle is held in a pair of pliers and inserted at an inconspicuous point. The easiest method, however, is to prick the stone with the point of a knife or needle. If the surface is brittle, and small bits come loose, then it is most likely amber.

Test methods

1. The looking and touching test

Professionals who are used to handling amber will in most cases discover a counterfeit even if they can't explain how. But because of constant improvements in imitating amber, even they will be fooled into starting work on a counterfeit if they don't conduct the necessary tests. It is important to remember that natural amber seldom has the symmetry and consistency in form, structure, and color associated with artificial amber. But some cases are more complex. It is often a good idea to bring a piece of genuine amber along to use as a comparison.

When natural amber jewelry ages, it is common to find cracks in its surface areas, and the regions around, for example, the holes of amber beads are usually slightly worn. This type of weathering usually does not occur with the artificial products.

2. The rubbing test

If amber is rubbed with a wool or silk fabric or, if nothing else is available, against a pant leg or a dress, it emits a slight scent of resin. However, this is also the case with certain artificial products such as Polybern as well. In addition, the amber becomes electrically charged from this treatment and will attract bits of paper.

3. The needle test

If the scent produced by the rubbing test is too faint, or if additional confirmation is sought, inserting a heated needle into the stone will result in a stronger aroma. The needle, held with a pair of pliers, should be inserted at the most inconspicuous point of the stone. In addition to natural amber, Polybern and amberoids smell like resin as well. Copal emits a slightly different resin scent, and the other imitations produce less agreeable smells. When heated or ignited, celluloid smells like camphor, casein smells like burnt milk, Bakelite smells sharp and pungent, and animal horn smells like burnt hair.

4. The water test

Amber floats in saturated salt water, certain plastics float in fresh water, and the other forms of artificial amber sink in salt water. This test, however, is useful only for artifacts that are not encased in, or combined with, other materials. The best way of conducting the water test is to use three standard drinking glasses. The first glass is filled with one tablespoon of salt, the second with two tablespoons, and the third with three. Amber will float in the third glass and sometimes in the second. Heavy plastic and glass will sink in all three glasses, while light plastic, such as polystyrene, will float even in the first glass.

5. The knife test

If a piece of amber is cut carefully, the outer surface layer will crack and crumble. The same is true for copal and amberoid. Bakelite tends to break into larger fragments. But shavings may be cut off the plastic materials with the ease of peeling a potato. Because this test may be difficult to conceal, a needle may be used instead of a knife in this case as well.

6. The ether test

A drop of ether on genuine amber does not produce an immediate reaction, although it should be wiped away quickly so it won't affect the luster. A drop of ether on a plastic imitation, however, will quickly dissolve the surface and make it granular. Even copal will soon gain a sticky surface if ether is applied.

7. The specific gravity test

Calculating the specific gravity of the material can be another solution. The volume may be calculated (as we all know from school) by lowering the object in water and measuring the rise in the surface level of the water. By dividing the object's weight out of water (in grams) by its volume (in cubic centimeters), its specific gravity is found. The result may later be compared to the natural specific gravities of amber. Professionals have additional testing methods available to them, such as measuring an object's refraction index or fluorescence. They may also employ chemical or spectral analysis. Amber is best separated from amberoid by using polarized light. Thus a jeweler with a polariscope should have no trouble in distinguishing between the two.

Caring for your amber

Jewelry made from natural amber becomes more beautiful the more it is worn. But the skin's secretions may with time discolor pieces such as necklaces and armbands. They should be cleaned as needed in lukewarm water, not hot water, with a little dishwashing liquid. Afterward, it is advisable to rub in a little vegetable oil. Because amber loses its luster when exposed to alcohol or ether, it should be protected from close contact with these or other solvents. It is important to remember that perfume and hairspray contain alcohol. If the amber jewelry should break, it can often be fixed with epoxy glue. For translucent amber, use cyanoacrylate glue (Super glue) instead so the joint cannot seen. If you are careful, it is possible to join the separated pieces though slow heating. Before doing this, you should rub the amber with amber oil, if available, or linseed oil. The safest way, of course, is to turn to an expert. Small surface blemishes may be gently filed or polished away.

If the string in an amber bead necklace needs to be replaced, remember to tie a knot between every bead. This will prevent them from rubbing against each other and from scattering all over the place if the string should break.

Amber artifacts should be stored in an airtight environment so as not to compound the corrosive effects of the air's oxygen. For this reason, amber objects are often stored in special exhibition cases. Still, much amber has been ruined over time. The deterioration usually begins with the object developing a network of tiny cracks.

Dominican and Mexican amber appear to be more volatile than their Baltic counterparts in this respect, and as a result, they age faster. One way of protecting amber from oxidation is to coat it with a thin layer of transparent polyester plastic. The layer is painted on and allowed to dry. But it may be difficult to feel the same way about your amber piece after doing this.

Furthermore, amber should not be exposed to sunlight for long periods of time. Exhibited artifacts should be protected with UV filters.

The holes of old genuine trading beads often show clear signs of wear.

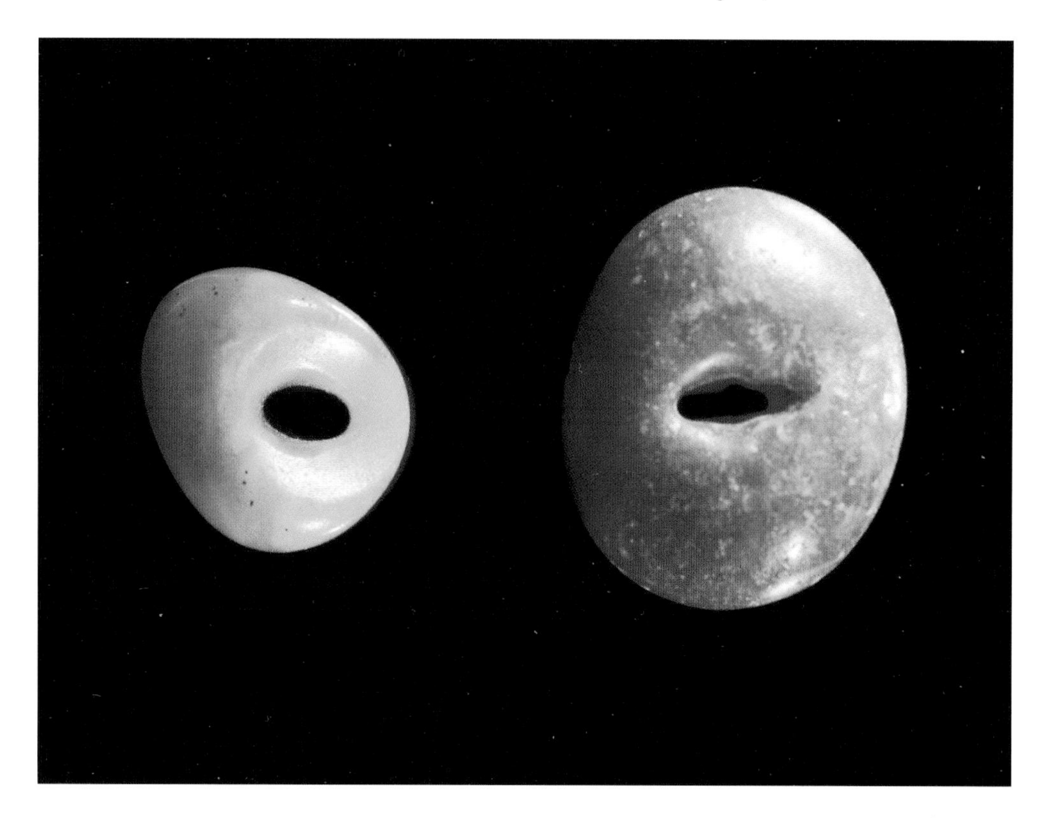

Facts about the more important kinds of amber

	AGE GEOLOGICAL PERIOD	DENSITY	HARDNESS	REFRACTION INDEX
Baltic amber				
Succinite	*Eocene–Oligocene*	*0.90–1.10*	*2.00–2.50*	*1.54*
Gedanite		*1.06–1.07*	*1.00–1.50*	
Beckerite				
Stantienite				
Glessite			*2.00*	
Rumanite	*Oligocene–Miocene*	*1.05–1.08*		*1.44*
Simetite	*Miocene*		*2.50*	
Dominican	*Miocene*	*1.05–1.08*	*1.50–2.00*	
Burmite (Burmese)	*Eocene*	*1.03–1.09*	*2.50–3.00*	*1.54*
Mexican	*Oligocene–Miocene*	*1.50–1.10*	*1.50–2.00*	*1.54*
Reconstituted amber, amberoid		*1.04–1.08;*		*1.54*
Fossilized copal	*Quaternary*	*1.03–1.10*		*1.54*

SUCCINIC ACID CONTENT	COMMON COLORS	COMMENTS
3–8%	*Yellow, brown, red*	*Most important kind of amber.*
Traces	*Pale yellow*	*Brittle, breaks easily. From Gedania, Latin for Gdansk.*
Traces	*Dark brown (opaque)*	*Soft yet tough. Origin presumably leguminous plant. Named after one of the owners of the firm Stantien & Becker, which developed the Baltic amber industry.*
Traces	*Dark, almost black*	*Reminiscent of beckerite but more brittle. Named after the other owner of Stantien & Becker. Called Schwarzharz.*
None	*Light brown*	*Made from resin of deciduous tree. Named after gles = "amber."*
1–5%	*Yellow-brown to deep brown*	*Most similar to succinite. Named after Romania.*
None	*Transparent, red, blue, green*	*Named after Simeto River, Sicily.*
None	*Pale yellow-brown, red, blue, green*	*High sulfur content. Origin: deciduous trees.*
Often	*clear*	
	Yellow-brown	

Geological time periods

EONS	ERA	PERIOD	EPOCH	MILLIONS OF YEARS AGO
	Cenozoic	*Quaternary*	*Holocene*	
			Pleistocene	*1.65*
		Teriary	*Pliocene*	*1.65–5*
			Miocene	*5–23*
			Oligocene	*23–35*
			Eocene	*35–57*
			Paleocene	*57–65*
	Mesozoic	*Cretaceous*		*65–146*
		Jurassic		*146–208*
		Triassic		*208–245*
	Paleozoic	*Permian*		*245–290*
		Carboniferous		*290–263*
		Devonian		*363–409*
		Silurian		*409–439*
				First land plants
		Ordovician		*439–510*
		Cambrian		*510–570*
	Proterozoic			*570–2,500*
Archean				*2,500–4,600*

CHARACTERISTIC LIFE-FORMS	AMBER FORMATIONS
Homo sapiens	*(Copal)*
Homo erectus	
Large predatory animals	
Whales, apes, graminivorous animals	*Dominican (20–40 mill. yrs.)*
Large mammals	*Succinite (30–50 mill. yrs.)*
Primitive horses, camels, large birds	
The first predatory animals and primates	
Dinosaurs become extinct	*Canadian*
Reptiles	*Siberian (80 mill. yrs.)*
Flowering plants (e.g., magnolias)	*Lebanese (125 mill. yrs.)*
Dinosaurs flourish	*Bornholm (170 mill. yrs.)*
Small mammals, birds	
Cycadeoids, ferns	
First dinosaurs	*Oldest known amber with fossil (225 mill. yrs.)*
Primitive mammals	
Forests of gymnosperms and ferns	
Reptiles	
Original conifers	
Amphibians, insects	*Oldest known amber occurrence (320 mill. yrs.)*
Large forests of gymnosperms and ferns	
Fishes, mussels	
Reef-forming coral	
Coral, starfish, octopuses	
Marine life-forms: protozoa, trilobites, algae	
Oldest known fossil	

Commercial classifications of succinite

TRANSLUCENT

Colorless to dark red-yellow. The "water-clear" type is very rare, the "yellow-clear" translucent type is most common.

OPAQUE

Fatty
Semitransparent with the appearance of "swirled honey," or translucent light yellow goose fat, as the German name for this amber type suggests.

Bastard
More or less cloudy, sometimes containing "cloud nebulas" in and of varying colors, such as white, egg-yolk yellow, brown-yellow, and red-brown.

Semibastard
Falls in between. Similar to bone amber, but may be polished for greater shine.

Bone (osseous)
Because of its soft and opaque nature it can't achieve the same shine as the semibastard. It is similar to bone or ivory in appearance.

Foamy (frothy)
Opaque, very light, and without shine.

Amber Artists circa 1550-1750

Stentzel Schmidt
Hired in 1563 as an amber craftsman by Duke Albert of Prussia.

Hans Klingenberger
Replaced Stentzel Schmidt.

Joachim Schöneberg
One of the leading craftsmen in Königsberg in the beginning of the seventeenth century.

Christoph Maucher
Seventeenth-century artist from Danzig, who completed many works for the Court in Brandenburg. His works include a beautiful casket that is now one of the foremost pieces in the Malbork collection.

Nicholas Turow
Maucher's partner. Was commissioned in 1677 to build a magnificent throne, with amber inserts, that was given to Holy Roman Emperor Leopold I of Austria.

Georg Schöneberg
Active in Königsberg until 1643. The foremost craftsman of the period. Signed his work Georgus Scriba at times.

Jacob Heise
Active during the middle of the seventeenth century.

Gottfred Wolfram
Employed at the court in Copenhagen. Initiated and worked on the Amber Room for seven years.

Ernst Schact and Gottfried Turau
Artists from Danzig. Continued Wolfram's work on the Amber Room between 1707 and 1711.

Jacob Dobberman
Employed at the court in Kassel, Prussia, in the early eighteenth century.

Michael Redlin
Active during the eighteenth century in Danzig.

Lorenz Spengler
Active in Britain and Denmark in the late eighteenth century. Produced portrait medallions and the famous candelabra in the Castle of Rosenberg.

Amber collections

Amber pieces and amber artifacts may be seen in the following museums, palaces, castles, and institutions.

SWEDEN

Amber Museum, Kämpinge
The museum houses amber pieces with inclusions from a variety of geological periods and parts of the world, as well as amber jewelry.

Castle of Gripsholm
The castle houses an amber cabinet from the eighteenth century decorated with ornate sculpture.

Castle of Skokloster
The museum's "silver chamber" houses amber drinking vessels, cases, crucifixes, eating utensils, and an amber pocket mirror, all from the seventeenth and eighteenth centuries.

Falsterbo Museum
The museum houses Parts of the Borgman collections and a picture in amber.

Historical Museum, Lund
The museum houses amber artifacts from the late Stone Age to the early Iron Age (beads, necklaces, buttons, animal figures, distaffs, etc.), as well as amber crucifixes of a later date.

Historical Museum, Stockholm
The museum houses amber artifacts from the Stone Age to the Middle Ages.

Institute of Paleontology, Uppsala University
The institute houses amber with animal inclusions.

Kulturen in Lund
The museum houses a miniature altar made of amber and ivory and silver-mounted pendants, as well as amber beads and dice found during a foundation dig in Lund.

Malmö Maritime Musem
The museum houses a boat model made of amber.

Malmö Museum
The museum houses the main portion of the Borgman collection.

Museum of the Middle Ages, Stockholm
The museum houses strings of amber beads and raw amber.

National Museum of Stockholm
The museum houses amber crucifixes, canisters, chalices, pitchers, saltcellars, board games, inkwells, and pictures from the seventeenth and eighteenth centuries.

Nordic Museum, Stockholm
The museum house amber cases, home altars, chalices, cigarette holders, and jewelry pieces.

Royal Academy of Science, Uppsala
The museum houses amber cases, mirror frames, board games, candleholders, crucifixes, folding knives, and reading monocles from the sixteenth to the eighteenth century. The collection comes primarily from the Castle of Drottningholm.

Royal Palace, Stockholm
The palace houses amber cases, jugs, and goblets from the sixteenth and seventeenth centuries.

Tobacco Museum of Stockholm
The museum houses a large collection of pipes, cigarette holders, and other tobacco accessories, some of them made from amber.

DENMARK

Castle of Rosenberg, Copenhagen
The castle houses several amber treasures, among other things a mirror by Spengler.

National Museum, Copenhagen
The museum houses a large collection of ancient amber.

Skive Museum, Skive

Zoological Museum, Copenhagen
The museum houses one of the world's finest collections of amber with animal inclusions.

OTHER COUNTRIES

Amber Museum, Palanga, Lithuania
Said to be the largest amber museum in the world.

Amber Museum, Puerto Plata, Dominican Republic

American Museum of Natural History, New York

Aquileia Museum, Aquileia, Italy

Archaeological Museum, Gdansk, Poland

Bezirksmuseum, Loma, Poland

British Museum, London

Castle of Malbork, Poland (3 miles southeast of Gdansk)
The castle is located in what used to be Marienburg, the headquarters for the old knights of the German Order. It exhibits amber from a variety of eras.

Deutsches Ivory Museum, Erbach, Germany
Around a fourth of the museum's exhibited pieces are made from amber. Most of the museum's pieces are from the sixteenth to the eighteenth century.

Grünes Gewölbe, Dresden, Germany

Hermitage, St. Petersburg, Russia

Kunsthistorisches Museum, Vienna

Metropolitan Museum of Art, New York

Museum of Fine Arts, Boston
The museum houses a handsome collection of European and Chinese amber art.

Museum am Löwentor, Stuttgart, Germany
The museum houses a large collection of natural amber specimens from around the world, many containing animal inclusions.

Museum für Naturkunde, Berlin
Among the museum's collections is the world's largest known piece of Baltic amber (succinite), weighing 21.5 pounds (9.8 kg).

Museum für Naturkunde, Stuttgart, Germany
The museum houses the current record holder for world's largest known piece of nonsuccinite amber, a specimen from Borneo in four pieces together weighing around 150 pounds (68 kg).

Museum of Paleontology, Berlin

Museum in Riebnitz-Damgarten (between Rostock and Stralsund), Germany

Museum Zeimi Pan (the Museum of the Earth), Warsaw
The museum has a permanent exhibit titled "Amber in Nature."

National Science Museum, Tokyo
The museum houses one of the world's largest known pieces of nonsuccinite amber, weighing 35 pounds (16 kg).

Natural History Museum, London
The museum houses a burmite stone weighing 33 pounds (15 kg), once considered the world's largest known piece of amber.

Palazzo Pitti, Florence, Italy

Schatzkammer der Residenz, Munich, Germany

Smithsonian Institution, Washington, D.C.

Staatliche Kunstsammlugen, Kassel, Germany

University Museum, Göttingen, Germany

Victoria and Albert Museum, London

Bibliography

BOOKS

Andersson, Leif. *Preparering av mineral och andra stenmaterial.* Sjöbo, 1982.

Bachofen-Echt, Adolf. *Der Bernstein und seine Einschlüsse.* Vienna, 1949.

Baltyku, Zloto. *Bärnsten: Guldet från Östersjön.* Warsaw, 1992.

Bauer, Max. *Precious Stones.* 2nd ed. London, 1904. Reprint, New York, 1968.

Berneking von Bock, Gisela. *Bernstein: Das Gold der Ostsee.* 1981.

Botfeldt, Knud. *Rav.* Copenhagen, 1980.

Geliad, S. *The Tears of the Heliades.* Moscow, 1991.

Gimbutas, Marija. *The Language of the Goddess.* New York, 1991.

Grabowska, Janina. *Polnischer Bernstein.* Warsaw, 1982.

Grimaldi, David A. *Amber: Window to the Past.* New York, 1996.

Hommerberg, Clas Peter Emil. *Bärnstenar.* Malmö, 1947.

Hunger, Rosa. *The Magic of Amber.* London, 1947.

Jensen, Jörgen. *Nordens guld.* Copenhagen, 1982.

Kosmowska-Ceranowicz, Barbara. *Spuren des Bernsteins.* 1991.

Kristensen, Frants. *Rav: Fra harpiks til smykke.* Copenhagen, 1986.

Krzeminska, Ewa, et al. *Les Fantômes de l'Ambre.* Neuchâtel, 1992.

Ley, Willy. *Dragons in Amber.* New York, 1951.

Ludwig, G. *Sonnensteine.* Berlin, 1988.

Museum of the World Ocean. *Amber and Fossils.* Vol. 1. Kaliningrad, 1995. With articles by Barbara Kosmowska-Ceranowicz and G. S. Kharin.

Pelka, Otto. *Bernstein.* Berlin, 1920.

Poinar, George O., Jr. *Life in Amber.* Stanford, California, 1992.

Poinar, George O., Jr., and Roberta Poinar. *The Quest for Life in Amber.* Reading, Massachusetts, 1994.

Reinecke, Rolf. *Gold des Meeres.* Rostock, 1986.

Rice, Patty C. *Amber: The Golden Gem of the Ages.* Rev. ed. New York, 1993. Extensive international bibliography.

Rodhe, Alfred. *Bernstein: Ein deutscher Werkstoff.* Berlin, 1937.

Rudal, Klaus. *Bernstein: Ein Schatz an unseren Küsten.* Husum, 1993.

Schlee, Dieter. *Bernstein-Raritäten.* Stuttgart, 1980.

——. *Bernstein-Neuigkeiten.* Stuttgart, 1980.

——. *Das Bernstein-Kabinett.* Stuttgart, 1990.

Stjernqvist, Berta, Curt W. Beck, and Jan Bergström. *Archaeological and Scientific Studies of Amber from the Swedish Iron Age* (Royal Society of Science and Humanities). Lund, 1994.

Traegårdh, Isaac. *Prövning i Historia rörande ön Glessaria.* 1752.

Webster, R. *Gems.* London, 1975.

Wikberg, Sven. *Från Bärnstenskusten och ordensriddarnas land.* Stockholm, 1931.

Sahlin, Carl. "Den skånsksa bärnstenen och dess tillgodogörande." In *Med hammare och fackla: Årsbok utgiven av Sankt Örjans.* Gille, 1936.

Wallin, Sigurd. Articles about the Uppsala Royal Science Society's amber objects in the society's annual publications of 1960 and 1962.

ARTICLES

B:son Flygare, Gösta. "Snidare och skalder." *Limhamniana,* 1969.

——. "Gammalt och nytt om bärnsten." *Limhamniana,* 1980.

Cano, Raul J., et al. "Bacillus DNA in Fossil Bees: An Ancient Symbiosis?" *Applied and Environmental Microbiology,* June 1994.

Journal of Baltic Studies, vol. 16, 1985. Articles by Curt W. Beck, George Poinar, Marija Gimbutas, and John M. Todd, among others.

Kristersson, Mikael. "Bärnsten." *Skånes Natur,* 1972.

Ley, Willy. "The Story of Amber." *Natural History,* May, 1938.

Ross, John F. "Treasured in Its Own Right, Amber Is a Golden Window of the Long Ago." *Smithsonian,* January 1993.

Photo credits

Bengt Almgren
72, 74

Leif Brost
11, 13, 14, 20, 21, 24-25, 27, 28, 30, 33, 35, 36, 37, 39, 55, 56 (top), 62 (left), 81, 95, 100, 112, 121

Ulf Bruxe
71, 73, 77, 88

Patrick Craig
26, 29, 38, 113

Tomas Grundström
10, 18, 19, 22 (bottom), 54, 56 (bottom), 61, 76, 79

Christer Johansson
62 (right), 68

Kjell Lindgren
60

Stig Ljung
cover, 78

Stanislaw Michta
82, 84, 85, 86

Rolf Reinicke
12, 22 (top), 63, 97

Bengt Serenander
15, 48, 50, 51, 53, 75, 103

Göran Strandberg
109

(Continued from page 57)
many parts of Japan. The most important find has been near the city of Kuji, north of Tokyo. This is also the site where the largest stones have been found. Japan might hold the world record in terms of amber size. It has been said that a thirteen-year-old boy found an amber boulder weighing 132 pounds (60 kg). For three days, he attempted to free it from the earth, but when other amber seekers began catching on and putting pressure on him, he became afraid of losing his claim to the stone and smashed it to pieces. But a stone weighing 35 pounds (16 kg), which may be seen in the National Science Museum in Tokyo, places Japan near the top of the list of contestants for world records. Another, found in 1927 and privately owned, weighs even more, close to 44 pounds (20 kg). Judging by its specific gravity, however, it contains other, less valuable materials than amber. (In more recent years, another world record contestant, from Sarawak on Borneo, has been brought forward. It is currently located in the Museum für Naturkunde in Stuttgart and is said to have weighed around 150 pounds (68 kg) before it was broken into four pieces—which still makes these the world's largest known for nonsuccinite amber.) It is believed that the oldest Japanese stones are around 110 million years old. They come in a variety of colors.

China's amber is located primarily in the Fu Shun area, north of Beijing. It dates from the Eocene period, around 60 million years ago.

Romanian amber (*rumanite*) is more closely related to succinite than the other variations. It is of roughly the same age, while Sicilian amber (*simetite* after the Simeto River) is of a more recent era. Unfortunately, the reserves of *rumanite* as well as *simetite* have practically been exhausted by now.

Slovakian or Czechslovakian amber has also been given a special name, *walchowit*. Said to be derived from the Cretaceous period and to be 100 million years old, it is wax-yellow and opaque.

In Spain amber may be found in the Basque region. A large find from the Cretaceous period has been made close to the capital city of Vitoria. It is said to be rich with inclusions.

Succinite has a few relatives in the Baltic area, but none of them are particularly valuable. The best-known kinds are *gedanite, beckerite, stantienite*, and *glessite*. More information on these and other types of amber is provided in the chart on pages 122–123.

Succinite amber can be classified in many ways according to structure, color, frequency of inclusions, and so on. Commercially, amber is divided into two main classes, "opaque" and "translucent," and these into hundreds of subclasses. The opaque stones are divided into five main subclasses; among others are the previously mentioned "amber stone," and a commonly highly valued variety called "bastard" (see page 126). The more natural distinction, dividing amber between "land amber" and "sea amber," should not be overlooked. The amber we find in the sea is generally the finest. It has rounder forms, with beautiful colors in the surface fractures. It is the fractures that cause the stones to shimmer. Natural land amber has a matte, corroded surface layer but may also be beautiful, especially when used to contrast with the shimmering sea amber. It is the oxygen of the atmosphere that causes the corrosion. Water protects amber from this effect—and delays the stones' aging.